The Nature of New Hampshire

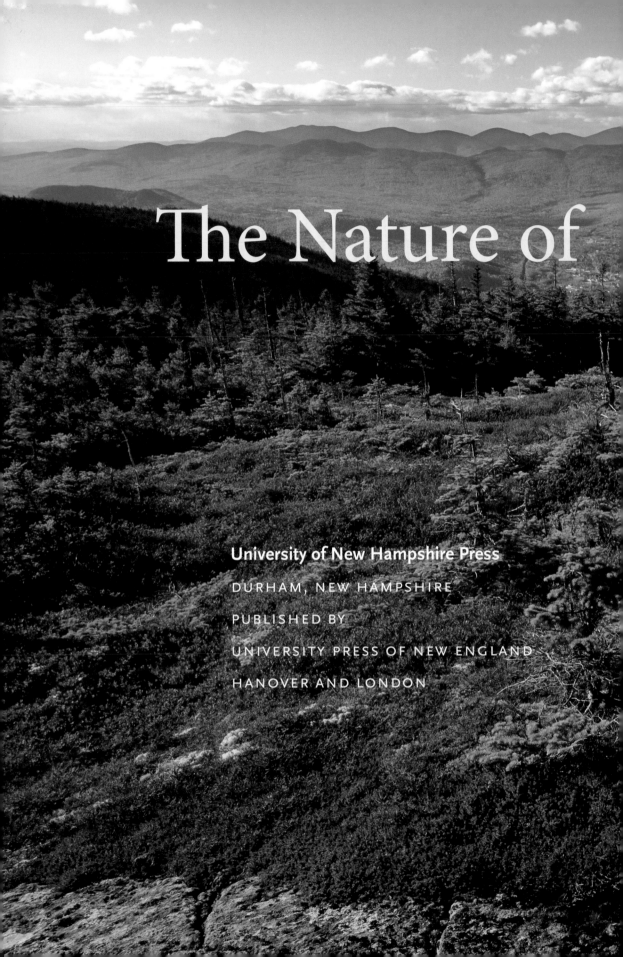

The Nature of

University of New Hampshire Press

DURHAM, NEW HAMPSHIRE

PUBLISHED BY

UNIVERSITY PRESS OF NEW ENGLAND

HANOVER AND LONDON

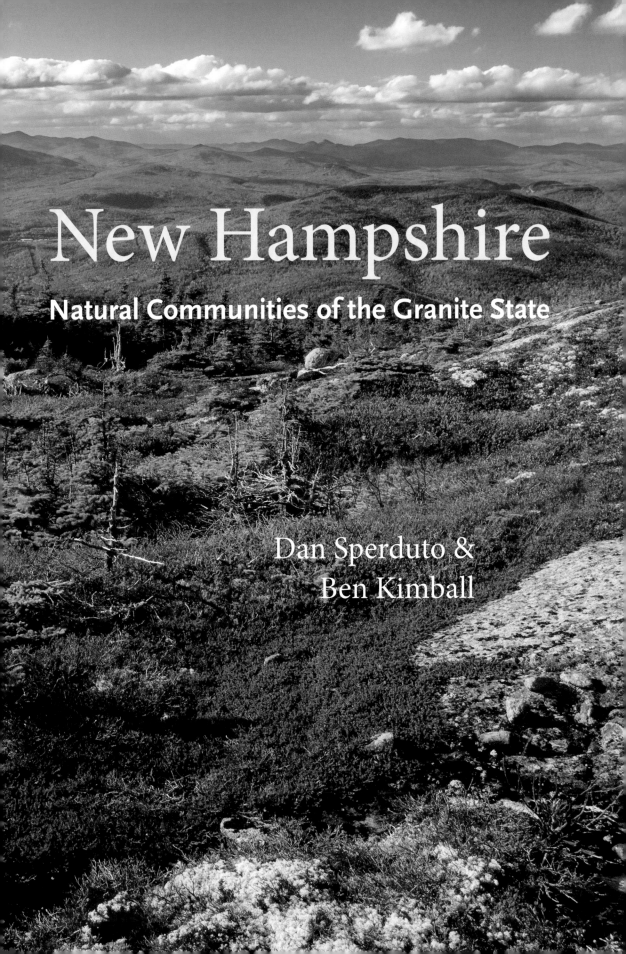

New Hampshire

Natural Communities of the Granite State

Dan Sperduto &
Ben Kimball

University of New Hampshire Press

Published by University Press of New England

www.upne.com

© 2011 University of New Hampshire

All rights reserved

Manufactured in Singapore

Designed by Katherine B. Kimball

Typeset in Minion by Passumpsic Publishing

For permission to reproduce any of the material in this book, contact
Permissions, University Press of New England, One Court Street,
Suite 250, Lebanon NH 03766; or visit www.upne.com

Library of Congress Cataloging-in-Publication Data

Sperduto, Daniel D.

The nature of New Hampshire : natural communities of the granite state / Dan Sperduto and
Ben Kimball. — 1st ed.

p. cm.

Includes bibliographical references and index.

ISBN 978-1-58465-898-6 (pbk. : alk. paper)

1. Biotic communities—New Hampshire. 2. Landforms—New Hampshire. 3. Ecological
regions—New Hampshire. 4. Plant communities—New Hampshire. 5. Natural history—New
Hampshire. I. Kimball, Ben. II. Title.

QH105.N4S64 2011

577.09742—dc22   2010039786

5 4 3 2 1

TITLE PAGE SPREAD: Shelburne-Moriah Ridge in the White Mountains region.

Contents

Foreword

Imagine the typical Granite State resident, suddenly transported to the pitcher's mound at Fenway Park in the middle of the seventh inning, staring down toward the batter at home plate. After the initial shock wears off, it's likely that person would know exactly where he or she is and, without turning to look, could describe the players on the field. Some could instinctively point over their right shoulders at the Green Monster, others might recognize the sound of a good hit by the crack of the bat on the ball. Drop those same residents into the middle of the Atlantic white cedar swamp in the northwest corner of Manchester and odds are they wouldn't have a clue about the natural community directly in front of them, never mind what they might see or hear in the woods behind them.

Most of us recognize a baseball game when we see one, and have at least a general understanding of the purpose and character of a baseball field. Too many, however, can't recognize a floodplain forest or, more importantly, appreciate why it exists and how it relates to adjacent swamps, marshes, and rivers. *The Nature of New Hampshire: Natural Communities of the Granite State* tells that story—and the stories of nearly two hundred other types of natural communities statewide.

This book is a clear-eyed and unsentimental accounting of the diversity of the state's natural communities, both common and rare, about which many of us are unabashedly sentimental. Rich descriptions of each community frame the context, chart the patterns of change and variability, and introduce us to many of the signature plant and animal species that are residents or key visitors to these habitats. Authors Sperduto and Kimball provide us with a new vocabulary—an enriched common language—for sharing an appreciation of the natural communities found throughout New Hampshire.

While not strictly a field guide, I expect many copies of this book to get wonderfully dirty in marshes or meadows. I suspect it will also prove a delightful resource for those who've climbed to an alpine summit, hiked through pine barrens, or slogged across a seacoast marsh; and for those who have sensed the special value of a place, want to understand it more deeply, and want to share that understanding with others. I dearly hope it will find a place in many classrooms.

I've read recently that the average young adult in America can identify one thousand corporate logos, but cannot name ten local species, let alone ten natural communities. This book—put to use in our forests, meadows, bogs, and fens—is part of the antidote for that dangerous ignorance. I imagine a day when a typical New Hampshire teenager might walk through that

Manchester swamp and know the cedar by the color of its leaves and the pattern of its bark. She might anticipate, without looking down, the tea-colored water and *Sphagnum* moss underfoot. Gazing at the deeply furrowed bark and blunted top of a black gum, she'd recognize a very old tree. Perhaps she'd even know from the tentative clip of hoof on rock that a white-tailed deer was coming down the trail. And then, walking up that trail herself, she'd marvel at the change from swamp to oak-pine woodland.

From the very beginnings of humanity, we have relied on the natural world for sustenance. As human societies have grown and developed, nature has supplied the raw materials. While those truths remain unchanged, too many people have lost touch with the beauty, solace, and perspective to be found there. We are gradually learning, however, how the unintended consequences of our actions can cause damage on both a local and a global scale.

Here's my hope: this book will be put to use by people who want to reconnect with their local natural communities and strengthen a shared purpose in stewarding the world around us. I believe that reconnection can help reinvigorate our will to better understand the natural world, and so fuel the will to conserve and restore it. And by preserving nature, we ultimately preserve ourselves.

Daryl Burtnett
Director,
The Nature Conservancy in New Hampshire

Acknowledgments

We are grateful for the contributions of many supportive partners, colleagues, and organizations who helped make this book possible.

Bill Nichols, ecologist at the New Hampshire Natural Heritage Bureau (NH Natural Heritage), provided critical input on myriad aspects of the project. Bill's work on *Natural Communities of New Hampshire* (Sperduto and Nichols 2004) informed much of the source material here, and he provided guidance on all classification revisions. His contributions to the final version of the manuscript were invaluable. Other NH Natural Heritage staff contributed significantly as well. Pete Bowman provided editorial review and ecological consultation. Don Kent helped with many aspects of the project, including administration and manuscript editing. Sara Cairns assisted with database support and maintenance of exemplary natural community information. Melissa Coppola consulted on botanical issues and provided comments on many community descriptions. Jeff Tash researched material on wildlife in alpine, rocky ground, and rich woods communities. Lionel Chute was an early champion of the project and provided the initial administrative work.

At the New Hampshire chapter of The Nature Conservancy, Daryl Burtnett has been a consistent supporter of the project, and we are especially thankful for his contribution of the book's foreword. Doug Bechtel provided project administration, editorial review, expert insight, and enthusiastic support throughout the project.

Emily Brunkhurst and Allison Briggaman with the New Hampshire Fish and Game Department provided foundational material for most of the wildlife sections. NH Natural Heritage and the Fish and Game Department's Nongame and Endangered Wildlife Program collaborate on maintaining data about the state's rare and threatened wildlife species.

Elizabeth Farnsworth drew all of the final illustrations of natural communities and ecological processes. We are grateful for her professional skill and endless patience with minor revisions. Marie Mead with the New Hampshire Bureau of Graphic Services worked tirelessly with all elements of the graphic design work, including the illustration annotations, species tables, and initial layout concepts. Her creativity and ability to meet tight deadlines are much appreciated. Dan Ostroth also provided consultation and administration assistance with the graphic design work.

Karen Bennett, Tom Lee, Scott Bailey, David Burdick, Colin Apse, Charlie Cogbill, and Tom Rawinski all carefully reviewed and proofread portions of the manuscript, and we are greatly appreciative of their wealth of knowledge and keen attention to detail. We also derived great benefit from the insights and expertise of many fellow ecologists and botanists over the years, including those from neighboring state heritage programs, NatureServe, and The Nature Conservancy. These colleagues include Mark Anderson, Lesley Sneddon, Stephanie Neid, Kathy Crowley, Larry Master, Andy Cutko, Sue Gawler, Ken Metzler, Julie Lundgren, Pat Swain, Greg Edinger, Eric Sorenson, Elizabeth Thompson, Bill Leak, Joann Hoy, Bill Brumback, Natalie Cleavitt, Brett Engstrom, and Peter Ellis.

Other people we wish to thank for their contributions and support include George Bald, Brad Simpkins, Phil Bryce, John Kanter, Charlie Bridges, Mike Marchand, David Wunsch, Holly Summers, Nicole Warren, David VanLuven, Kayden Will, and Molly Sperduto.

We would especially like to thank the University Press of New England for their patience, support, and advice during the production phase of the book, in addition to their roles in copyediting, layout, and publication.

Finally, NH Natural Heritage is indebted to the New Hampshire Conservation License Plate Program for funding staff time to write the book. We thank the program for its commitment to protecting and conserving the state's native plants and natural communities. We hope that this book serves the dual purpose of informing the people of New Hampshire about the state's precious natural heritage and contributing to informed, sciencebased conservation action.

About the New Hampshire Natural Heritage Bureau

To administer the New Hampshire Native Plant Protection Act (RSA 217-A), the New Hampshire Natural Heritage Bureau collects and analyzes data on the status, location, and distribution of native plant species and natural communities in the state. In addition, the bureau develops and implements measures for the protection, conservation, enhancement, and management of native New Hampshire plants and natural communities, determines which plant species are threatened and endangered, and acts as an information resource program to assist and advise state and local agencies, and the public. The New Hampshire Natural Heritage Bureau is a part of the Department of Resources and Economic Development, Division of Forests and Lands.

The New Hampshire Conservation License Plate Program

Purchase of a Conservation License Plate promotes, protects, and invests in New Hampshire's natural, cultural, and historic resources. You can purchase a Conservation License Plate at your municipal office, whether you are registering your car for the first time or renewing an existing registration. In addition, you can purchase a gift certificate for someone else to use when registering or renewing his or her car. For details about the Conservation License Plate Program, visit: www.mooseplate.com.

The Nature of New Hampshire

Introduction

Natural Communities

Natural communities are recurring assemblages of plants and animals found in particular physical environments. Three characteristics distinguish natural communities: (1) plant species composition, (2) vegetation structure (for example, forest, shrubland, or marsh), and (3) a specific combination of physical conditions (including water, light, nutrient levels, and climate). Each natural community type occurs in specific settings in the landscape, such as wind-exposed rocky summits at high elevations, or muddy coastal river shores flooded daily by tides. Natural community types vary with changes in physical settings, resulting in predictable patterns across the landscape.

New Hampshire is a mosaic of natural communities, encompassing habitats as varied as alpine meadows, riverbanks, forests, tidal marshes, ponds, and cliffs. Communities range from common and widespread types that cover hundreds or thousands of acres across broad areas of the state, to uncommon or rare types that are small and restricted to a specific part of the state — such as the White Mountains or Seacoast region. Across much of the landscape, a few forest types form a matrix, with other natural community types occurring as patches embedded within that matrix.

Protecting Biodiversity

Biodiversity is the variety and variability of all living organisms. It includes whole organisms, their genes, the natural communities in which they live, and the complex interactions among and between organisms and their physical environment. Natural levels of biodiversity may be very high, as in tropical regions with favorable growing conditions and many different species. In deserts and arctic regions, where conditions are harsh and few species can survive, biodiversity can be very low. The biodiversity in a given area decreases when species suffer local extinctions, when invasive species displace native species, and when the natural habitats that support local species are fragmented or destroyed. Unique components of biodiversity, such as species or natural communities that have a limited distribution, are a focal point for conservation efforts.

Biodiversity has important value. Some values come from direct uses, including recreation, medicine, and forest products. Other values derive from indirect uses, such as soil building, erosion control, fire prevention, water quality improvement, contaminant absorption, flood reduction, crop

FACING PAGE: A typical second-growth forest in Hopkinton.

pollination, and pest control. Finally, biodiversity has important non-use values. People attach value to species or places simply by knowing of their existence, even if they may never see or experience these resources firsthand.

Species interactions are complex and poorly understood. Species with little apparent value may play a critical role in the survival of beneficial species, or in the suppression of harmful species. The loss of a single species—and the resulting disturbance of a natural community—can have unpredictable consequences.

Persistent components of healthy natural communities and ecosystems are called biological legacies. One type of biological legacy is topsoil, the layer of mineral earth that contains a large quantity of organic material from the growth, death, and decomposition of plants. Other types of biological legacies include seed banks, coarse woody debris, and soil nutrients, as well as subterranean fauna. Biological legacies take many years to develop, and are diminished when natural processes are disrupted.

Long-term protection of New Hampshire's species, natural communities, and ecological processes requires multiple approaches. Natural communities are by definition assemblages of multiple species; protecting a community therefore provides protection for many individual species. Consequently, if we conserve an adequate number of viable examples of each natural community type, we can protect the majority of New Hampshire's species. This approach is a "coarse filter" for protecting biodiversity. The coarse filter can miss some important species, and should be augmented with a "fine-filter" approach that focuses on rare species. By locating populations of these species, then protecting the natural communities where they occur, we can complement coarse-filter conservation and protect the full range of biodiversity.

Coarse woody debris in an old-growth forest at Pisgah State Park.

In many instances, protection of one natural community may require protection of adjacent communities or systems. All communities, regardless of size, interact to some extent with their surroundings. Processes such as erosion, windfall, fire, and nutrient accumulation at a given site are affected by the context of the broader landscape. In addition, many animal species require access to a combination of communities in close proximity to satisfy their food and shelter requirements throughout the year. Biodiversity conservation is therefore enhanced when protected areas include a variety of adjacent or connected communities.

The size of protected areas makes a difference; bigger is better. Long-term community viability increases as the size of protected areas increases, and larger areas support certain wide-ranging animals that would not occur in smaller areas. Edge effects such as invasive species establishment are also reduced in larger, protected areas. The foundation for successful biodiversity protection is a network of representative, high-quality examples of a region's natural community types, and by extension, their constituent species and underlying ecological processes.

Classifying Natural Communities

Classifying natural communities enables ecologists, land managers, and others to communicate effectively about a site's biodiversity and ecological context, thereby informing land stewardship and conservation planning. Natural community classification provides a framework for objective evaluation and comparison among sites. Factors for comparison include rarity, relative size, and ecological integrity, in both state and regional contexts (see appendix 1 for more details). As landscape categories that share physical and biological characteristics important to many species, natural communities focus management and conservation attention on practical ecological units that serve as effective coarse filters for biodiversity.

The New Hampshire Natural Heritage Bureau (NH Natural Heritage) developed the classification of natural community types presented in this book. That effort grew out of work originally undertaken in the northeastern United States by The Nature Conservancy. Other state and provincial natural heritage programs in the United States and Canada developed similar classifications.

The New Hampshire natural community classification is based on more than twenty years of ecological research by NH Natural Heritage, The Nature Conservancy, and others. NH Natural Heritage compiled and analyzed the data to define natural community types by applying ordination and other quantitative classification techniques, supplemented by reviews of the scientific literature. NH Natural Heritage updates and improves the classification as new data are gathered. See the NH Natural Heritage website for future revisions to the New Hampshire natural community classification (www.nhnaturalheritage.org).

In this book, the names of natural community types appear in **bold** at first mention, and when otherwise necessary to avoid confusion. These names generally begin with the dominant or most characteristic plant species, and may include the name of a specific landscape feature or vegetative structure typical of the community type. For example, **black gum - red maple basin swamp** refers to a basin swamp with abundant black gum and red maple in the tree canopy. A hyphen offset by spaces indicates different species or characteristic groups of plants. For example, in **pitch pine - heath swamp**, pitch pine is in the tree canopy and the heaths are in the shrub layer. Hyphens without spaces are an inherent part of some species names, such as twig-rush, or are word modifiers. Slashes represent and/or combinations;

for example, **herbaceous riverbank/floodplain** indicates that the community can occur on riverbanks or floodplains, or both.

To define vegetation structure, the New Hampshire natural community classification uses percent cover of the component plant forms. We apply the following structural categories when naming and describing communities.

Forest: 60 to 100 percent tree cover; shrubs, herbs, and non-vascular plants may be present at any cover value.

Woodland: 25 to 60 percent tree cover; shrubs, herbs, and non-vascular plants may be present at any cover value.

Shrub thicket: 60 to 100 percent shrub cover and less than 25 percent tree cover; herbs and non-vascular plants may be present at any cover.

Shrubland: 25 to 60 percent shrub cover and less than 25 percent tree cover; herbs and non-vascular plants may be present at any cover.

Herbaceous: greater than 25 percent herb cover; trees and shrubs may be present but each with a cover of less than 25 percent; non-vascular plants may be present at any cover value.

Non-vascular: greater than 25 percent non-vascular plant cover; trees, shrubs, and herbs may be present with a cover of less than 25 percent.

Sparsely vegetated: less than 25 percent total vegetation cover.

The New Hampshire natural community classification differs from other classifications by considering nutrient regime, water source, and geomorphic setting. Differences in floristic composition indicate the combined effect of these factors. In addition, the New Hampshire classification provides details about ecological conditions and processes that explain the distribution of biological diversity across the landscape.

NH Natural Heritage bases its natural communities on plant species, the structural layers that these species collectively form, and the specific physical environment. Several species of trees are indicators of subtle differences in their environments. A number of natural communities can be distinguished based largely on trees, and in some instances, a difference in tree composition is the main difference between two community types. Conversely, trees such as white pine or red maple are so broadly adapted that their presence does not necessarily indicate specific site conditions. As with trees, certain species of shrubs and herbs have a narrow range of ecological tolerances and are found in only a few types of natural communities. Other shrub and herb species, such as black huckleberry and rough goldenrod, are generalists similar to white pine and red maple. Human-related disturbances can increase the importance of physiological tolerance to stress and, as a result, shift species composition. In some settings, for example, tree canopy composition may be more influenced by cutting or other human disturbance than other environmental factors.

Just as some species are further classified by subspecies, natural communities that exhibit distinctive variation within the type can be described by individual variants. Variants are characterized by relatively minor vegetation differences and either minor or major soil differences. They may include a shift in dominant tree species where the understory vegetation remains identical, or a simple shift in abundance of one or more species. Variants are described in a more technical version of the natural community classification system on the NH Natural Heritage website (www.nhnaturalheritage.org).

Exemplary Natural Communities

NH Natural Heritage evaluates the ecological significance of individual natural communities and assigns a quality rank. Quality ranks are a measure of the ecological integrity of a community relative to other examples of that particular type, based on size, ecological condition, and landscape context. Natural disturbances such as ice storms, blowdowns, and fires are a critical component of natural system processes and do not diminish a community occurrence ranking. NH Natural Heritage designates all rare natural community types, and high-quality examples of more common community types, as exemplary. Exemplary natural communities represent the best remaining examples of New Hampshire's biological diversity. NH Natural Heritage identifies and tracks exemplary natural community occurrences to inform conservation decisions.

Natural Community Systems

Natural community systems are associations of natural communities in the landscape, linked by a common set of characteristics, such as landform, frequency of flooding, soil, and nutrient regime. Each major group of communities in this book typically corresponds to one or two natural community systems. In chapter 3, for example, Acadian spruce - fir forests correspond with the lowland spruce - fir forest/swamp system and the high-elevation spruce - fir forest system. Systems are useful units for a variety of conservation planning and mapping purposes. New Hampshire's natural community systems are described on the NH Natural Heritage website (www.nhnaturalheritage.org).

The Biophysical Environment

The distribution of plants and communities is a function of climate, geology, soils, and community dynamics. Temperature and moisture vary with latitude, elevation, aspect, and microtopography. Bedrock and surface deposits influence landform, water flow and storage, and soil. Soils affect water and nutrient availability. Disturbance, competition, and succession are important community dynamics. Vegetation's adaptation to its biophysical environment determines the composition and structure of natural communities.

Climate

Climate is the prevailing weather conditions of a region over time. It is a composite of temperature, air pressure, humidity, precipitation, solar energy, cloudiness, and winds, averaged over a series of years. Annual and seasonal variation in solar energy produces three global climate zones: tropical, temperate, and polar. Temperature and precipitation variability lead to smaller climatic regions within these zones. Plants are adapted to the temperature, light, water, and nutrients of climate regions. Less directly, climate influences long-term ecosystem processes such as bedrock weathering, soil development, and nutrient cycling and storage.

Critical climatic variables for plants include temperature extremes, moisture levels, and growing-season length. All plants have tolerance ranges of temperature, moisture, and solar energy, above or below which they cannot survive. Over time, many plant species have developed mechanisms for extending their tolerance range. For example, adaptations to cold include a seasonal hardening of tissue by woody plants, leaf fall by deciduous plants, and overwintering as seeds by annual species.

As used in this book, a climate region's growing season is the time between killing frosts. The vegetation of a climate region is adapted to its growing season. For example, alpine tundra plants complete their life cycles within 100 days. By contrast, plants of southern New Hampshire hardwood forests are adapted to a growing season of 150 days or more.

New Hampshire's climate is cool-temperate, although the state's weather is more variable than other temperate regions of North America. The state experiences hot summers, cold winters, drought, heavy rain, snow, thunderstorms, hurricanes, nor'easter storms, and tornados. New Hampshire's climatic variability is due to its geographic location and topographical diversity. Air masses move west to east across the state, with continental air masses hot in the summer and cold in the winter. In addition, the northeastern United States is a convergence zone for major storm tracks. New Hampshire receives storms that originate in the Pacific Ocean, Rocky Mountains, Gulf of Mexico, South Atlantic Ocean, and arctic regions. Precipitation is plentiful, averaging 42 inches per year. Winter storms are typically stronger and more frequent than summer storms because polar air masses produce

Winter conditions at Lonesome Lake in the White Mountains.

strong temperature contrasts with moist air masses from the Gulf of Mexico and the South Atlantic Ocean.

Intense weather systems such as hurricanes, thunderstorms, and nor'-easters cause wind and other storm-related damage to New Hampshire forests. Impacts from these storms include tree blowdown, wind and ice pruning, and root damage. Storm impacts are modified by local topography (aspect, elevation) and vegetation characteristics (tree species, height, and strength). The most destructive storms are rare, but they do have a significant and long-lasting impact on species composition. Powerful hurricanes, such as the one in 1938, have come from the south and resulted in greatest damage to forests in southeastern, southern, and central New Hampshire, with progressively less destructive impacts to the north and inland.

New Hampshire's north–south orientation and coastal locale also contribute to the diversity of its climate. The state spans 2.5 degrees of latitude, enough to create different climates from north to south. For example, a mix of spruce, fir, and northern hardwoods dominates forests at the 1,000-foot elevation in northern New Hampshire, whereas oaks dominate at the same elevation in the southern part of the state. The ocean influences climate by moderating air temperatures, producing milder winters and cooler summers near the coast. When free of ice, Lake Winnipesaukee is large enough to influence the climate within several miles of its shore (although to a lesser extent than the ocean).

Mountains and other topographic features also modify local weather and climate. Air masses cool as they rise over mountains, increasing cloud cover and precipitation. The White Mountains and other high-elevation areas experience colder temperatures, extended snowpack, and shorter growing sea-

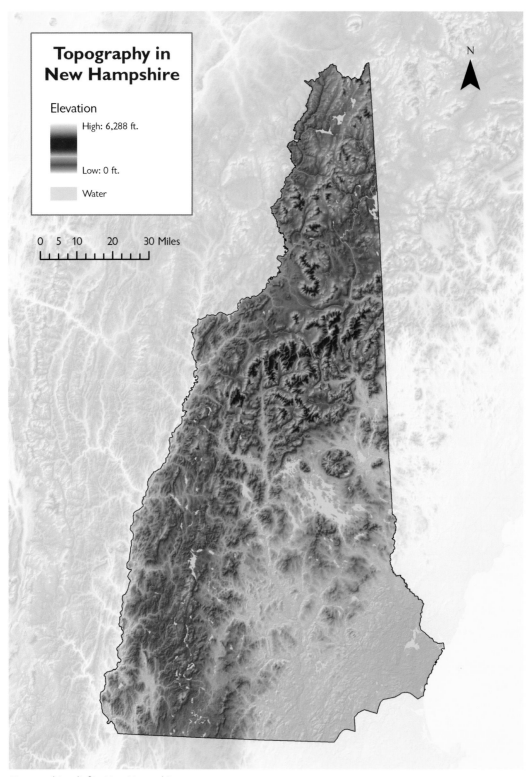

Topography in
New Hampshire

Elevation

High: 6,288 ft.

Low: 0 ft.

Water

0 5 10   20   30 Miles

N

Topographic relief in New Hampshire.

sons than lowland areas. In addition, rocks and vegetation at high elevation intercept clouds and fog, with condensation adding to the available moisture. Precipitation is generally greater on the windward side (the one facing the oncoming wind) than the leeward side of mountains.

Elevation directly affects temperature, contributing to the state's weather and climate variability. The elevation range in the state is greater than in any other state or province in northeastern North America, starting from sea level and topping out at 6,288 feet on Mount Washington. On average, an elevation change of 320 feet produces a change in temperature equal to that caused by a shift of 1 degree of latitude. Thus, the climate at the top of Mount Washington is equivalent to that of Iceland or Greenland, about 20 degrees to the north. Elevation and landform also affect wind speed. Wind speed increases as air moves up and over mountains. Notches experience increased wind speeds as air is forced through a confined area.

Cold, dense air settles in topographic depressions as the ground cools down at night. As a result, early morning temperatures are often colder in valleys, lake basins, and other low-lying areas compared to adjacent higher ground. The cold air shortens the growing season, an effect that is exacerbated if the depression collects water. Conversely, a topographic depression with a seep may influence an area's microclimate by extending the growing season. Seeps, the surface outlets for groundwater, maintain a relatively constant soil temperature, and can reduce or prevent freezing.

Solar radiation varies seasonally, with changes in day length and in the angle of the sun above the horizon. Slope, aspect, and microtopography influence local climate by affecting the amount of solar radiation received. For example, steep, north-facing slopes receive less solar energy than steep, south-facing slopes, and are colder as a result. Deep ravines in the mountains often form shaded, cold-air pockets with late-melting ice and snow. An extreme example is Ice Gulch in Randolph, which often contains ice until August and supports alpine plants at elevations well below the alpine zone. Conversely, steep, south-facing slopes in the White Mountains are warm enough to support red oak, a tree species otherwise restricted to lower elevations or southern New Hampshire. Temperature and moisture conditions can vary with very small changes in topography, such as between the top and bottom of a boulder or hummock, or from one side of a tree to the other.

Geologic History and Landforms

New Hampshire's beautiful landscape of mountains, hills, valleys, lakes, rivers, terraces, and coastline is the product of hundreds of millions of years of rock formation, upheaval, and erosion. Repeated glacial advances and retreats eroded bedrock and modified the distribution of the loose rock fragments and sediments collectively known as surface deposits. Together, bedrock and surface deposits determine the shape of the land. Landforms affect the flow and storage of water, and modify local climate conditions. In turn, water flow and storage, along with local climate conditions, influence soil characteristics and vegetation. Finally, soil characteristics strongly influence plant species composition and cover, two features that are important in distinguishing natural communities.

BEDROCK

Bedrock determines the shape of the landscape, yet is exposed on only a small percentage of New Hampshire's surface area. Close examination of the state's bedrock reveals a complex mosaic of different rock types. Three bands of bedrock, oriented from northeast to southwest, occupy north, central, and southeastern New Hampshire. The southeastern band of rock is the oldest, up to 650 million years old. Bedrock in the northern part of the state is of more recent origin, 520 to 475 million years old; central region bedrock is even younger, 455 to 365 million years old. Within each band are smaller, more recent igneous intrusions formed by molten material. These younger igneous rocks are associated with the break-up of the Pangaean supercontinent and opening of the present-day Atlantic Ocean basin. A warm, wet climate prevailed during most of New Hampshire's 650-million-year geologic history, enabling erosion to produce large masses of surface deposits prior to glaciation.

GLACIATION

Glaciers first covered New Hampshire 1.6 million years ago, marking the first in a series of glacial advances and retreats during the Pleistocene epoch. The last glaciation period ended 10,000 to 11,000 years ago, after the retreat of the Wisconsin Age ice sheet. Continental glaciers removed less than 200 feet of the land surface in most parts of the state, but gouged lake basins from weakened areas of bedrock, plucked rocks from fractured ledges, smoothed hills and ridges, and transported and re-deposited surface material. At high elevations, alpine glaciers further eroded the landscape, scooping steep-walled, U-shaped ravines called cirques into mountainsides. Prior to glaciation, the rugged White Mountains likely resembled the smooth, rolling skyline of the present-day Smoky Mountains of North Carolina.

SURFACE DEPOSITS

Surface deposits—mixes of boulders, stones, cobble, gravel, sand, silt, and clay—cover most of New Hampshire's bedrock. Three types of surface deposits occur in New Hampshire: ice-deposited glacial till, water-deposited sediments, and other deposits formed or modified after glaciation. Surface deposits strongly influence soils and plant communities.

Glacial Till   The majority of New Hampshire's surface deposits are glacial till. Till is a mix of boulders, stones, gravel, sand, silt, and clay once trapped within or beneath a glacier. These sediments were subsequently left behind on the land surface as the ice melted away. There are two main types of till: basal till and ablation till.

Basal till, also known as compact till, was compressed by the weight and force of the glacier above. It is dense and impedes water infiltration and root penetration. Some basal till is left over from previous glaciations. Landscapes covered by basal till tend to have a smooth or even surface and are common throughout the state, particularly on the lower slopes of mountains and hills. Basal till often forms the base of drumlins, elliptical hills with their steeper face on the side opposite the glacier's direction of travel; in New Hampshire these faces are typically on the southeastern side. Drumlins are most common in central and southern parts of the state.

Ablation till, or loose till, consists of sediments formerly inside, next to, in front of, or on top of the glacier. Less dense than basal till, ablation till offers less resistance to water infiltration and root penetration. Ablation tills are common throughout the state, particularly on middle and upper landscape positions and as a surface layer on top of basal tills. Landscapes covered by ablation till tend to have a choppier, more uneven land surface.

Water-Deposited Sediments   New Hampshire's lowlands are comprised of sediments deposited in rivers, streams, lakes, and marine environments during glacial retreat. Many of these aquatic environments are now smaller, or have dried up entirely. This type of sedimentation continues to this day, but is less extensive than during the glacial period.

The particle size of deposited sediments is determined by water velocity and turbulence. Cobbles, stones, and gravels settle to the streambed in turbulent or relatively high-velocity conditions. Particles of sand, silt, and clay are smaller; they settle only as floods subside and stream velocity decreases. Changes in flow volume and velocity produce layers of different-sized sediments called stratified deposits. For example, a glacial stream that diminishes in flow volume and velocity over time would produce a series of layers, with cobble at the bottom, and subsequent layers of gravel, sand, and silt. Sediments set down in stream or river environments, called fluvial or alluvial deposits, are mostly coarse, ranging from cobble to silt. By contrast, lakebed sediments consist mostly of silt and fine sand. Near-shore marine deposits consist of silt and clay.

Fluvial deposits include ice-contact sediments deposited inside or along

Glaciation in New Hampshire

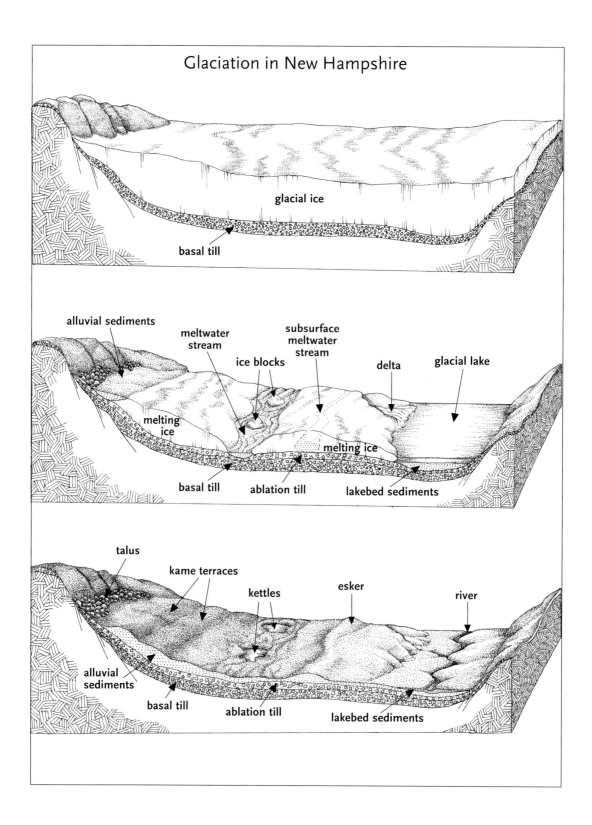

glacial ice

basal till

alluvial sediments

meltwater stream

subsurface meltwater stream

ice blocks

delta

glacial lake

melting ice

melting ice

basal till

ablation till

lakebed sediments

talus

kame terraces

kettles

esker

river

alluvial sediments

basal till

ablation till

lakebed sediments

the margins of a melting glacier, and sediments deposited beyond the terminus of a glacier. Ice-contact deposit landforms include eskers and kames. Eskers are long, narrow, sinuous ridges of stratified sand and gravel that formed in a stream channel inside or along the margin of a melting glacier. Kames are irregularly shaped ridges, hills, or terraces of stratified sediments formed in meltwater pockets within a glacier, or in streams along the edge of a glacier. Kames that formed along the margin of a glacier are called kame terraces. These typically flat-topped and steep-sided terraces are a common landscape feature in today's river and stream valleys.

Melting glaciers fed rivers that transported and deposited large amounts of coarse and fine sediments downstream. The sediments settled and formed outwash deposits in major river valleys, on glacial lake plains such as the Ossipee region, and in river delta areas such as the Concord Heights. Sometimes blocks of ice were stranded and partially buried by outwash deposits. These blocks eventually melted, forming kettle holes in the landscape. Lakes and ponds now occupy many of these depressions.

As the glaciers receded, wind-blown dunes formed as meltwater levels dropped and exposed sandy deposits along the margins of glacial lakes and on outwash plains. Today, active dunes occur only along the coast, and along a few sandy river terraces of the Merrimack River.

Floodplains are flat areas bordering a river. During flood events, water spreads over the floodplain, slows, and deposits gravel, sand, silt, and clay. River terraces are historic floodplains formed during glacial meltdown periods (when rivers were larger), or more recently due to channel migration or downcutting (the process by which a river erodes its bed). Sand and silt are the most common deposits forming floodplains and river terraces, although coarser outwash or kame materials may occur adjacent to existing rivers. Oxbows are narrow, often curved depressions in a floodplain, remnants of

FACING PAGE: *Glaciation and landscape features in New Hampshire.* The upper panel shows a deep continental glacier in place, overlying basal till. A nunatak (unglaciated mountain summit) is visible at the left. The middle panel shows a degrading ice sheet with various meltwater features. Plucking has deposited talus at the base of the mountain. An ice-contact zone to the right of the toe of the mountain will later form a kame terrace (see lower panel). Large, abandoned ice blocks to the right of this ice will later form kettle-hole ponds. An esker is being formed along the channel of a subsurface meltwater stream. A stream running into the meltwater lake at the right is depositing deltaic beds. At the bottom, ablation till is being deposited by the melting glacier onto basal till. The lower panel shows landforms left after the ice melted completely. The mountain is now fully exposed, with talus at its toe. Kame terraces are present in the former ice-contact zone. In the level plain to the right of the terraces, two kettle ponds have formed. An esker ridge occurs farther to the right. Deltaic beds are visible at the shore of the former lake, which is now a flat plain being downcut by a stream.

former river channels left behind after a river changed course. Ponds sometimes occupy oxbows.

Glacial lakebed deposits consist of clay, silt, and fine sand that settled to the bottom of historic glacial lakes, temporary bodies of water created behind ice or debris dams. The deposits often form varves, alternating layers of sediments deposited over the seasons: coarser material in the summer, finer in the winter. Coarser sands and gravels also were deposited along the margins of such lakes by incoming rivers and streams. Eventually the ice and debris dams failed, the glacial lakes drained, and glacial meltwater streams deposited material on top of the glacial lakebed deposits. Deep, coarser fluvial sediments underlain by finer, varved lake sediments are exposed on steep terrace slopes of the existing Connecticut and Merrimack rivers.

Marine sediments characterize some coastal lowlands and valley bottoms of New Hampshire. Much of the landscape that is now above sea level in the coastal region was underwater when the glaciers were receding, as the land had not yet rebounded from the glaciers' weight. Fine silt and clay sediments were deposited on the seabed, and in many places coarse outwash deposits covered the marine deposits. Because marine silts and clays impede water infiltration and root penetration, wetlands are common in areas where marine deposits are near the ground surface.

Other Surface Deposits   Small areas of New Hampshire's landscape are comprised of mineral materials that originated from sources other than water or glacial deposition. Talus is a sloping mass of rock fragments usually found at the base of a cliff. In most instances, the parent cliff still rises above the talus but, in some cases, the cliff has completely eroded away. Felsenmeer is comprised of exposed rock fragments overlying parent bedrock; it forms where frost and ice crack bedrock and create boulder fields. Felsenmeer depth varies with climate, vegetation, and rock type. In New Hampshire, felsenmeer occurs in alpine areas. Landslides are avalanches of rock, soil, and other debris on steep mountain slopes. A landslide typically eliminates all vegetation in its path and creates a debris cone of mixed rubble at its base. Talus and felsenmeer maintain natural communities in an early successional stage, while landslides periodically reset natural communities to earlier successional stages.

Soils

New Hampshire's soils developed from bedrock and surface deposits following the retreat of glaciers. The percolation of water, physical and chemical breakdown of rocks and sediments, and the actions of plants and animals all modify the parent material to form soil. Soils provide plants with a sub-

Dry Brook in Franconia Notch rarely runs dry. All told, New Hampshire has thousands of miles of rivers and streams.

strate for growth and a source of water and nutrients. As biophysical environments, soils encompass mineral and organic matter, fungi, bacteria, invertebrates, plant roots, and other organisms. Soil properties vary with climate, landform, topographic setting, surface material, and water movement. Factors associated with soils that affect plants and natural communities include water flow and drainage, soil horizons, bedrock geology, and nutrient cycling.

WATER FLOW AND SOIL DRAINAGE

Water levels influence both the development of soil and the vegetation that grows in the soil. The amount of water available in the root zone is a significant factor in determining natural community composition. Some plants have developed specific adaptations in response to periods of inundation, significant water-level fluctuations, or xeric soil conditions.

Soil-water fluctuation is a function of water input, retention, and output. Water inputs include precipitation, surface and near-surface inflow, and groundwater discharge. Water retention is a function of landscape position, topography, and soil porosity. Water outputs include evapotranspiration, surface and near-surface runoff, and groundwater recharge (see illustration, p. 16). Losses from evapotranspiration peak during the growing season, and are greater in warmer climates than colder ones. In addition, evapotranspiration is a function of vegetation type and leaf surface area. Growing season evapotranspiration is greater in deciduous plants than evergreens, and greater in communities with high total leaf surface area, such as forests, than in more sparsely vegetated communities.

Landscape position affects water input and retention. Precipitation drains

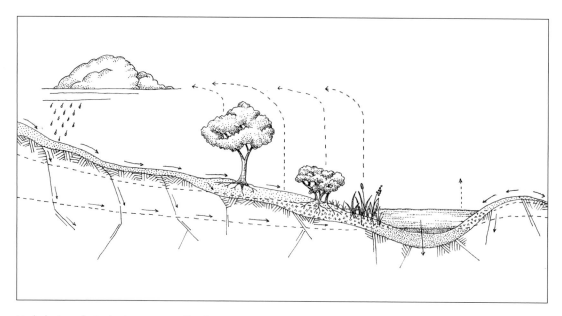

Hydrologic cycle. Bedrock is shown along the base of the landscape. Fracturing in the bedrock allows water to flow to the lower groundwater aquifer. Shallow soils cover the bedrock. Streams deliver water both to the groundwater layer and to the wetland shore and pond at the right. Deeper organic soils are shown with shading beneath the pond. Losses of water due to evaporation and transpiration are shown with dotted lines and arrows. The two dashed lines running horizontally reflect the mean and highest levels of the water table.

rapidly from steep or convex landscapes such as ridges or hillsides. Water then collects in concave and level areas along rivers and streams, and in valley bottoms and wetland basins. Water may be temporarily stored in vegetation and soils, or as groundwater.

Soil drainage is a function of topographic setting and physical soil properties such as texture, density, and porosity. Water remains at or above the surface for most or all of the year in topographic depressions with restricted outflow or on soils with low porosity. Conversely, water remains well below the surface on steep slopes or in soils with high porosity. The Natural Resources Conservation Service categorizes the former as poorly or very poorly drained, and the latter as excessively or somewhat excessively well drained. Intermediate drainage classes include well drained, moderately well drained, and somewhat poorly drained.

SOIL PROFILES, PROCESSES, AND TYPES

Soil types vary with differences in climate, landforms, topographic setting, water flow, parent materials, and biological activity. Localized chemical and physical changes result in the development of layers in the soil called horizons. Each soil horizon varies in thickness, color, texture, structure, consistency, porosity, acidity, and composition.

There are four major soil horizons. The surface layer is called the O horizon, comprised of loose leaves, woody material, and partly decomposed organic debris. Below this layer is the A horizon, consisting of organic and mineral material mixed together by the combined action of soil organisms, such as nematodes and insects, and the downward movement of organic fragments from the O horizon. The A horizon is often darker than other layers due to a higher content of organic matter. This is also the layer of maximum leaching, the transport of nutrients and organic humus to lower soil layers by percolating water. The B horizon accumulates clay minerals and organic matter leached from the O and A horizons. It is more varied in color (usually lighter than the A horizon), and characterized by a blocky or prismatic structure. Mineral material beneath the B horizon that has not been modified by soil-forming processes is called the C horizon. Organic and inorganic acids enhance leaching of soil nutrients from surface layers. As a result, upper soil layers become more nutrient deficient over time, contributing to distinct soil horizons.

Climate, water level, and nutrient accumulation affect the character of soil horizons. For example, leaf litter decomposes more slowly in New Hampshire's colder or wetter regions. As a result, O horizons in these areas are relatively thick, and the percolation of acidic water leaches fine sediments and organic matter out of the top layers of soil. A gray, coarse-textured E horizon develops in place of the A horizon, or between the A and B horizons. In warmer parts of the state, the thickness of the O horizon diminishes and the thickness of the A horizon increases. The warmer climate and longer growing seasons facilitate decomposition of organic matter, and more thorough mixing with mineral soil. Soils comprised of recently deposited alluvium, such as those on river floodplains, have indistinct horizons.

Many wetland soils, particularly those characterized by sustained saturation or inundation, can develop a deep O horizon of peat or muck. Peat is an accumulation of partially decayed vegetation fragments, while muck consists primarily of humus, highly decomposed organic matter. When squeezed, a fistful of muck feels denser and greasier than peat. Seasonal saturation or inundation may produce mineral soils. The recurring water-level fluctuations promote leaching, hence the A horizon may be thin or entirely absent. The extent of water-level fluctuation is evidenced by mottling, the appearance of colored splotches resulting from the oxidation of iron and manganese.

BEDROCK AND PLANT NUTRIENTS

Most nitrogen used by plants originates from the atmosphere. Other plant nutrients, such as calcium, magnesium, phosphorous, and potassium, originate from bedrock or mineral fragments. Chemical and physical weathering breaks complex mineral compounds into simple ionic forms that reside in soil water. These nutrients may attach to mineral or organic particles, leach out of the soil to streams or groundwater, or be taken up by plants. Nutrient availability to plants is determined by the mineral composition and weathering rate of the parent material.

Factors which affect weathering include the proportion of exposed surface area and the extent of fracturing. Climate also affects weathering. In cold climates, cycles of freeze and thaw expose more rock surfaces. In warmer, wetter climates, precipitation accelerates chemical weathering.

In New Hampshire, the nutrient supply typically is more limited than in other regions of North America. Most rocks have low concentrations of calcium, magnesium, phosphorus, and potassium, and a high resistance to weathering. In addition, calcium not taken up by plants is readily lost to leaching. Nonetheless, elevated levels of calcium and other nutrients do occur occasionally, and these areas support many uncommon or rare plants and natural communities.

Geologists differentiate rocks by origin, mineral composition, and crystalline structure. Igneous rocks form from the cooling and solidification of molten material. Sedimentary rocks form from consolidation of sediments deposited by wind or water. Metamorphic rocks form from the alteration and recrystallization of existing igneous or sedimentary rocks by high temperature and pressure. In New Hampshire, all bedrock is igneous or metamorphic; there is no sedimentary bedrock in the state.

Igneous rocks are comprised of a combination of light- and dark-colored minerals distinguished by their chemical composition. Light-colored minerals such as quartz and feldspar are composed primarily of silica, contain low concentrations of important nutrients such as calcium, and are resistant to weathering. In contrast, dark-colored minerals such as pyroxenes and amphiboles can have a higher concentration of nutrients and weather more rapidly.

Igneous rocks are categorized by silica content. Felsic rocks contain the most silica and have little calcium. Felsic rocks weather slowly to produce acidic, nutrient-poor soils, and are very common in New Hampshire. Intermediate igneous rocks have less silica and more calcium than felsic rocks; they weather more rapidly, frequently yielding soils that are less acidic and more nutrient-rich than soils derived from felsic rocks. In New Hampshire, intermediate igneous rocks are not widespread but often occur in ring dikes, crescent- or ring-shaped rock formations associated with ancient volcanoes. The Pawtuckaway Mountains, Ossipee Mountains, and Cape Horn are good examples of ring dikes containing intermediate igneous rocks. Mafic rocks have even less silica, higher concentrations of iron and magnesium, and often higher concentrations of calcium than felsic or intermediate rocks. Most mafic igneous rocks in New Hampshire weather slowly. Some produce soils with relatively high calcium content and low acidity. Mafic igneous rocks are uncommon and scattered in the state, but where they do occur they support important calcium-rich habitats.

Metamorphic rocks are very common in New Hampshire. Most are high in silica and low in calcium. They weather slowly and yield acidic, nutrient-

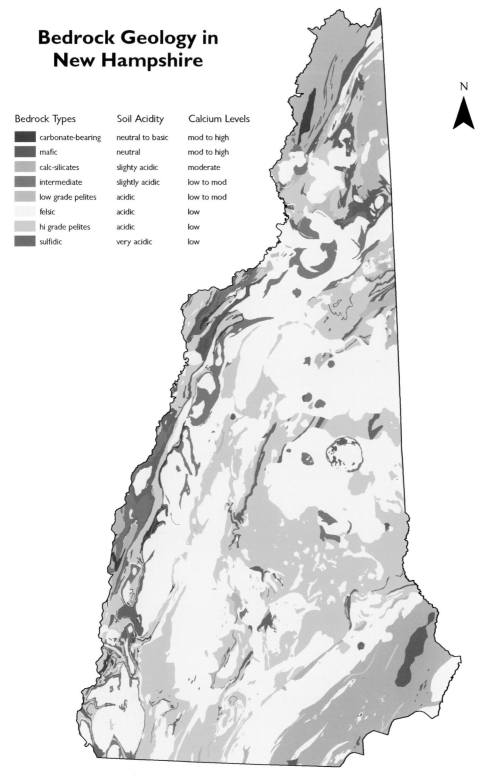

Bedrock Geology in New Hampshire

Bedrock Types	Soil Acidity	Calcium Levels
carbonate-bearing	neutral to basic	mod to high
mafic	neutral	mod to high
calc-silicates	slighty acidic	moderate
intermediate	slightly acidic	low to mod
low grade pelites	acidic	low to mod
felsic	acidic	low
hi grade pelites	acidic	low
sulfidic	very acidic	low

N

Based on: Bailey, S. W. 2000. Ecological setting of the northern forest—geologic and edaphic factors influencing susceptibility to environmental change. In R. Birdsey, J. Hom, and R. Mickler (eds) *Responses of Northern U.S. Forests to Environmental Change*. Springer-Verlag Ecological Studies Series 139:27–49.

poor soils. A few metamorphic rock types have higher calcium concentrations and weather more rapidly, producing calcium-rich soils that are less acidic than most soils in the state. Calcium-rich metamorphic rocks are limited to the North Country, a narrow band along the Connecticut River, and bands in the Seacoast region.

New Hampshire contains no limestone, a sedimentary rock comprised primarily of calcium carbonate, though it does occur in localized areas in other parts of New England. Limestone is more frequent elsewhere.

SOIL FERTILITY AND NUTRIENT CYCLING

Soil fertility exerts a significant influence on natural community type. Nutrients originate from bedrock and the atmosphere, and cycle between soils, plants, and animals. Soil fertility is the rate at which nutrients become available to plants, rather than the total amount of nutrients. That rate depends on the quality of the soil's organic matter, pH, nutrient forms, and weathering.

Plants obtain most of their nutrients from soils, but only indirectly from soil organic matter. Organic matter contains nutrients bound in complex organic compounds that plants cannot access. Microbes (mostly fungi and bacteria) and other soil organisms decompose organic matter into accessible forms in a process called mineralization.

For rapid mineralization to occur, these organisms require high-quality organic matter, easily digestible forms of carbon, and moist, well-aerated environments with circumneutral pH (6.0–7.9). High-quality organic matter contains relatively high concentrations of nitrogen and other nutrients, and low concentrations of slow-decomposing carbon compounds. Plants that produce high-quality organic matter, such as sugar maple, ash, and dogwood, are adapted to fertile soils and occur on rich sites. Factors that contribute to the development of rich sites include calcium-rich bedrock or till, and mesic topographic settings that accumulate organic matter.

The nutrients calcium, magnesium, and potassium exist as positively charged base cations in many New Hampshire soils. Base cation availability is limited by bedrock and soil sources, and the abundance of other cations such as aluminum and hydrogen, which displace calcium, magnesium, and potassium from available storage sites on soil particles. Plants rapidly absorb base cations not attached to soil particles, and any excess nutrients are leached from soil into streams and groundwater.

The availability of the most limiting nutrient controls the growth rate and total biomass of plants in a natural community. Nitrogen is the limiting nutrient in many upland and forested swamp communities. Nitrogen or phosphorus may be the limiting nutrients in other types of wetlands. Major groups of communities — such as forests, marshes, and peatlands — can differ substantially in terms of nitrogen and phosphorus availability and, as a result, in their plant productivity and biomass.

Community Dynamics

LONG-TERM VEGETATION CHANGE

Since the retreat of glacial ice, New Hampshire's vegetation has shifted in response to changes in climatic, hydrologic, soil, and disturbance patterns. Plants migrated from glacial refugia in surprisingly disparate regions, including the Arctic, Europe, now-submerged portions of the Atlantic coastal plain, the southern Appalachian Mountains, and the Midwest.

Fossil evidence preserved in glacial lakebed sediments along the Connecticut River indicates that the first plants to arrive after the final glacial retreat were tundra lichens, mosses, herbs, and shrubs. Most of these species require open, disturbed, or calcium-rich soils, indicating that the early postglacial landscape had different climate and soil conditions than today. Most of the tundra colonizers are now absent from New Hampshire, replaced by trees and other vegetation, except for relict populations in alpine and subalpine habitats on high mountain peaks.

Tree species arrived at different times in New Hampshire, reflecting a changing climate and differences in dispersal rates, migration routes, geographic barriers, and distances from glacial refugia. The attributes of each species' life history, such as dispersal mode, competitive ability, and climatic tolerance, determined the rate of establishment. The most rapid period of establishment occurred during the first several thousand years following the glacial retreat. Conifers and aspen, species with wind-dispersed seeds, arrived first. Spruce, larch, and red or jack pine (pollen of the latter two species cannot be distinguished from each other) arrived about 10,000 years ago. American elm and white pine followed within a thousand years, and eastern hemlock arrived about 8,000 years ago. American beech, hickory, and American chestnut arrived about 6,500, 5,000, and 2,000 years ago, respectively.

DISTURBANCE, COMPETITION, AND SUCCESSION

Wind, floods, droughts, fires, ice, and extreme temperatures periodically disturb many natural communities. Disturbance kills or damages vegetation, modifies site conditions, and shifts competitive balance and direct succession. Many species and communities depend on a certain intensity and frequency of disturbance to persist. For example, herbaceous riverbank species persist because periodic scouring eliminates woody plants that would otherwise dominate that part of the floodplain. Early successional plant assemblages maintained by natural disturbances, such as scoured riverbanks, constitute distinct natural communities.

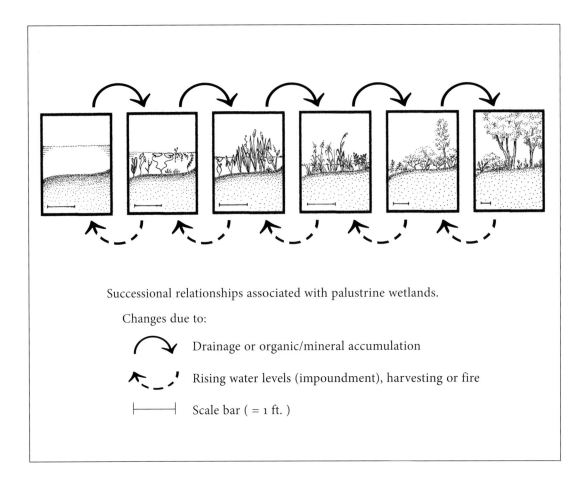

Successional relationships associated with palustrine wetlands.

Changes due to:

Drainage or organic/mineral accumulation

Rising water levels (impoundment), harvesting or fire

Scale bar (= 1 ft.)

Wetland succession in palustrine habitat. This wetland sequence progresses from open water (left), to aquatic plants, emergent plants, wet meadow, shrub swamp, and forested swamp (right). Soil depths remain the same, while water levels drop.

Biotic disturbance agents include herbivores and pathogens. Moose and deer eat a variety of plants, but prefer certain species. Gypsy moths and spruce budworms exhibit cyclic population outbreaks that can dramatically affect forest communities. Gypsy moths defoliate deciduous tree species, while spruce budworms target balsam fir, but may also damage spruce, hemlock, and pine. Familiar pathogens include chestnut blight fungus, which largely eliminated American chestnut from our forests, and Dutch elm disease, which caused a catastrophic die-off of American elm in all of the habitats it grew in, from wetlands to cities.

Species compete for light, nutrients, and water. When resources are unlimited, the species that grows most rapidly is typically more successful. When a resource is limited, the plant that is most efficient or most tolerant of the stress imparted by resource limitation usually prevails. For example, red oak attains its greatest size on mesic, moderately rich sites where sugar maple is abundant, but only dominates on nutrient-poor sites too dry for sugar maple.

Succession is the replacement of one community by another, usually but

Philbrick-Cricenti Bog in New London formed in a glacial kettle hole. This small pool is all that remains of the original pond's surface.

not always progressing to a more stable community type. Primary succession is the initial colonization of unvegetated areas, while secondary succession is the sequence of vegetative changes following disturbance. One form of primary succession, xerarch succession, occurs in upland landscapes. In this process, shade-intolerant pioneer species such as lichens and mosses initially colonize bare ground or rock, promoting soil accumulation or enrichment. Herbs, shrubs, and trees then arrive, eventually culminating in woodland or forest.

Hydrarch succession is a form of secondary plant succession, starting in relatively shallow water and culminating in a forested swamp. A shallow pond with submersed and floating-leaved plants accumulates organic matter and sediments, enabling the establishment of marsh vegetation. As organic matter and sediments continue to accumulate, shrubs colonize the marsh, and ultimately a forested swamp develops. The specific sequence of changes in vegetation depends on topography and hydrology. In wetlands with organic soils, a progression from fen to bog to peat swamp is typical in very poorly drained topographic depressions with limited seasonal water-level fluctuations. In better-drained wetlands with more seasonal water fluctuations, the sequence typically involves a progression from pond to marsh to mineral swamp. A disturbance — such as flooding of the wetland by beavers — sets back hydrarch succession.

FOLLOWING SPREAD: *Major differences in setting, hydrologic condition, and vegetation between types of peatlands, marshes, and swamps.* The wetlands on the left occur along a sluggish stream or pond with inlets and outlets and significant seasonal water level fluctuations. The wetlands at right have less variable water levels.

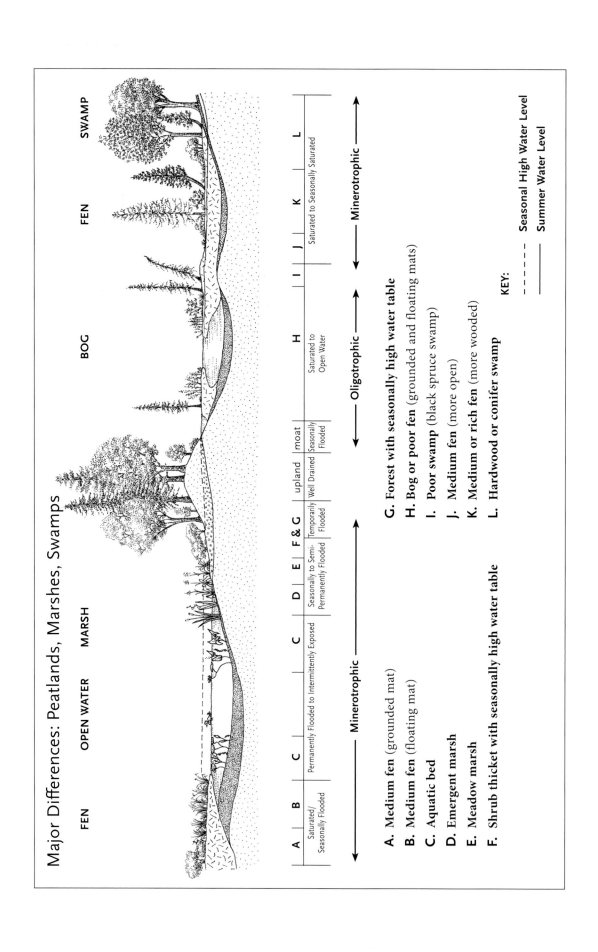

Major Differences: Peatlands, Marshes, Swamps

FEN   OPEN WATER   MARSH   BOG   FEN   SWAMP

A	B	C	D	E	F & G	upland	moat	H	I	J	K	L
Saturated/ Seasonally Flooded		Permanently Flooded to Intermittently Exposed	Seasonally to Semi-Permanently Flooded		Temporarily Flooded	Well Drained	Seasonally Flooded	Saturated to Open Water				Saturated to Seasonally Saturated

⟵ Minerotrophic ⟶   ⟵ Oligotrophic ⟶   ⟵ Minerotrophic ⟶

A. Medium fen (grounded mat)
B. Medium fen (floating mat)
C. Aquatic bed
D. Emergent marsh
E. Meadow marsh
F. Shrub thicket with seasonally high water table

G. Forest with seasonally high water table
H. Bog or poor fen (grounded and floating mats)
I. Poor swamp (black spruce swamp)
J. Medium fen (more open)
K. Medium or rich fen (more wooded)
L. Hardwood or conifer swamp

KEY:
– – – – – Seasonal High Water Level
————— Summer Water Level

SUCCESSIONAL PATHWAYS AMONG PALUSTRINE WETLANDS

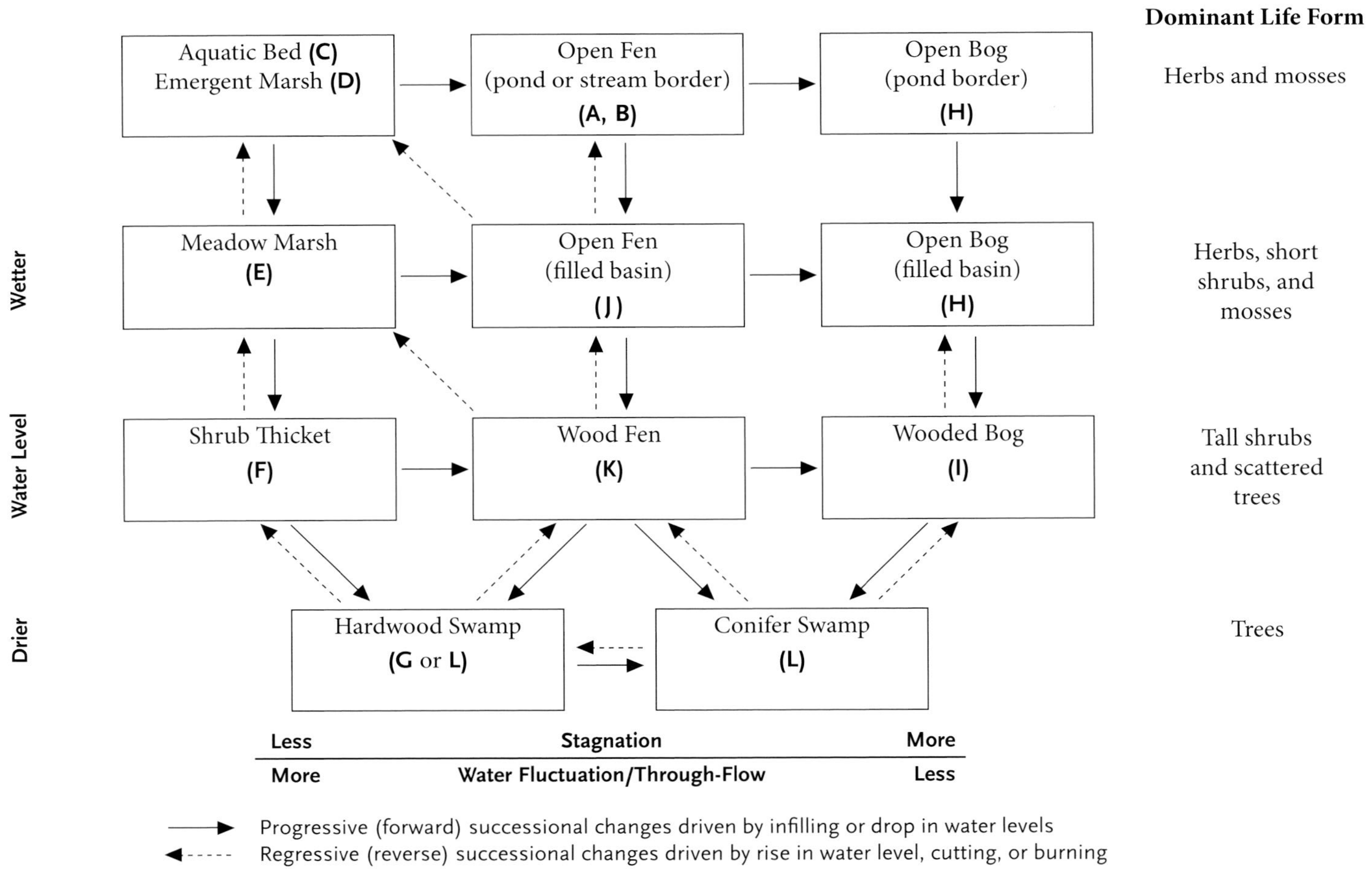

Dominant Life Form

Herbs and mosses

Herbs, short shrubs, and mosses

Tall shrubs and scattered trees

Trees

Aquatic Bed (C) Emergent Marsh (D)	Open Fen (pond or stream border) (A, B)	Open Bog (pond border) (H)	
Meadow Marsh (E)	Open Fen (filled basin) (J)	Open Bog (filled basin) (H)	
Shrub Thicket (F)	Wood Fen (K)	Wooded Bog (I)	
Hardwood Swamp (G or L)	Conifer Swamp (L)		

Wetter — **Water Level** — **Drier**

Less		More
More	Water Fluctuation/Through-Flow	Less
	Stagnation	

Progressive (forward) successional changes driven by infilling or drop in water levels

Regressive (reverse) successional changes driven by rise in water level, cutting, or burning

RARITY

Some natural communities are scarce in New Hampshire because the regions of the state in which they occur lie at the edge of their climatic range. For example, enriched Appalachian oak forest communities and their characteristic suites of understory species reach their northern climatic limit in the southeastern part of the state. Within New Hampshire, Appalachian oak forests are restricted to warm, south-facing hills. At the other climatic extreme, the state's alpine communities are limited to cold climates on the highest White Mountain summits. In contrast, some communities are rare because they contain a unique combination of plants with widely disparate geographic distributions. For example, **inland Atlantic white cedar swamps** contain a mix of coastal plain and boreal plants.

Other natural communities are rare because their physical habitat is rare. For instance, coastal sand dune communities can only occur where an adequate supply of sand exists and currents, waves, and wind cause sand to accumulate. In New England, this combination is unusual north of Massachusetts and, in New Hampshire, occurs only in Hampton and Seabrook. Calcium-enriched fens are another example of a natural community limited by physical habitat. Such fens require the convergence of calcium-rich bedrock and strong groundwater seepage, a rare combination in New Hampshire.

The scarcity of some New Hampshire natural communities is related to a dependence on disturbance for community maintenance. For example, the state's distribution of dry sandy soils could support more **pitch pine - scrub oak woodlands**, commonly known as pine barrens, than presently exist. Pine barrens require periodic fires for their maintenance and regeneration, and the plant and animal species found in these ecosystems are uniquely adapted to withstand fire. Without periodic burning, species intolerant of fire displace pine barren species. Fire suppression policies have curtailed burning to protect forest-product resources and real-estate development, allowing fire-intolerant species to become more prevalent. See appendix 2 for the state rarity rank of each natural community type.

HUMAN EFFECTS ON NATURAL COMMUNITIES

Land conversion, agriculture, timber harvest, recreation, and flood regulation directly affect New Hampshire's natural communities. Other human activities — such as habitat fragmentation, fire suppression, the introduction of non-native species, and air, water, and soil pollution — alter the composition and processes of natural communities in less obvious ways.

Humans are a natural part of the environment, but we have an effect on the landscape disproportionate to that of other species. Conversion of forest lands to agriculture and other uses began with Native Americans. In those times, cutting and burning were typically restricted to areas close to villages and traditional hunting grounds. When local resources diminished, villages were moved to other areas where productive soils allowed farming to complement hunting and gathering. New Hampshire's Seacoast region and the Connecticut River Valley were notably important areas for Native Americans.

Forest species composition and succession shifted in response to European settlement. Conversion of forest land to agriculture intensified, with grazing and plowing leading to increased soil erosion and edge effects on forest communities. Timber harvesting also increased. Intensive cutting in the White Mountains contributed to massive brush fires and caused large-scale soil erosion that silted major rivers and changed the nature of aquatic communities downstream. Log-drive scouring and dam construction further affected stream communities. Public outcry led to the creation of the White Mountain National Forest and the adoption of more informed land management practices.

All New Hampshire natural communities have been affected to some degree by human disturbance, although the effects on species and ecosystems are not evenly distributed. Development disproportionately affects rare species and natural communities. For example, the farmers who settled the region were attracted to rich floodplain soils and converted many lowland forests to crop fields. Today, New Hampshire's lowland forests occupy only a small part of their historic extent. Floodplain forests that have been recovered from agriculture, subjected to ditching, or that exist near dams do not reflect their original character.

Even broad-scale disturbances affect certain species, natural communities, and regions more than others. For instance, unchecked global climate change will shift the geographical distribution or species composition of many natural communities, and eliminate or greatly diminish species at climatic extremes, such as in alpine tundra communities. In addition, climate change will affect the frequency and intensity of certain disturbance regimes such as storms or floods. Another example involves the atmospheric deposition of nitrogen and other elements, also known as acid rain. Acid rain alters the availability of key soil nutrients to plants, long-term forest productivity, and frost susceptibility in certain trees.

Ecological Patterns and Regions in New Hampshire

Plant Distribution Patterns

Most New Hampshire plant species have broad geographic ranges centered in one of four different North American climate regions: arctic-alpine, boreal, temperate, and coastal plain. Some plants occupy only particular portions of a climate region. Other plants might not fit a specific region at all; these include species found throughout North America or the northern hemisphere, non-native plants, and plants with unique or unusual distribution patterns.

ARCTIC-ALPINE

Plants in New Hampshire's arctic-alpine region are restricted to alpine and subalpine peaks above 4,000 feet in elevation, with a few occurring on lower elevation cliffs, in ravines, and along rivers in the mountains. Most of these plants also occur in other alpine areas of northeastern North America, but are separated from their primary ranges in northern Canada. Examples of arctic-alpine region species include Bigelow's sedge, alpine bilberry, and highland rush. A few species are endemic to northeastern North America, and one, dwarf cinquefoil, occurs only in the White Mountains.

BOREAL

The boreal region encompasses the great northern conifer forest, extending from Alaska to eastern Canada (similar boreal forests occupy northern portions of Europe and Asia). In New Hampshire, boreal region species occur from the White Mountains northward, and in peatlands throughout the state. Boreal region plants include balsam fir, black spruce, paper birch, larch, and quaking aspen. Numerous species in New Hampshire are restricted to the southeastern portion of the North American boreal forests, or occupy the transition zone between boreal and eastern deciduous forests, including red spruce, red pine, northern white cedar, sheep laurel, and rhodora.

TEMPERATE

The eastern temperate region is located south of the boreal forest and east of the Great Plains, an area coincident with eastern deciduous forest. In New Hampshire, temperate region species occur throughout the state, below 2,500 feet elevation in forests and wetlands. Most New Hampshire species with temperate distributions occupy particular sub-regions or portions of the eastern deciduous forest. Hemlock, yellow birch, white pine, sugar maple, American beech, red maple, and black cherry occur throughout New Hampshire. Oaks, hickories, and dogwoods, representative Appalachian or central hardwood species, occur mainly in central and southern parts of the state.

COASTAL PLAIN

The coastal plain occupies a band 10 to 100 miles wide along the Atlantic and Gulf coasts. In New Hampshire, coastal plain species occur in the Seacoast region, and at low elevations in the Merrimack River Valley and the Lakes Region. Most are wetland or sand plain species, including Atlantic white cedar, dwarf huckleberry, sweet pepperbush, beach grass, golden heather, and Virginia chain fern. Many of New Hampshire's coastal plain region plants are rare or uncommon.

Types of Natural Communities in New Hampshire

New Hampshire has eight distinct biophysical natural community groupings. Each group is described briefly here, and discussed in detail in a later chapter.

ALPINE AND SUBALPINE

Alpine and subalpine communities occupy cold, wind-exposed summits and upper slopes, mostly in the White Mountains, above 4,000 feet elevation. The communities typically support dwarf shrubs and sedges with arctic-alpine distributions. The soils are shallow, rocky, and mostly well drained. The growing season is short, especially where snow lingers into late spring. Many of the plants in alpine and subalpine natural communities are rare in New Hampshire.

ROCKY GROUND

Rocky ground communities occupy areas of exposed bedrock in upper landscape positions. Plant cover is low, and includes scattered spruce, pine, and oak trees, heath shrubs, herbs, and lichens of boreal or temperate distribution. Rocky ground natural communities include outcrop areas on summits and ridgelines, cliffs, and talus slopes. They are dry or seasonally dry and remain open due to recurring fire, steepness, or regular rockfall. Rocky ground communities occur throughout the state below 4,000 feet elevation, and are most abundant in mountainous areas.

FORESTS

Forest communities occupy upland sites, and consist of moderate to dense canopies of trees above shrub and herb layers. Forests cover more than 80 percent of the state. Temperate and boreal plants dominate New Hampshire forests. Elevation, latitude, topography, soils, and prevailing disturbance regimes such as wind and fire all affect forest community type.

PEATLANDS

Peatland communities occur statewide and include bogs and fens, wetlands that typically occupy very poorly drained basins. Peatlands are saturated year-round, and water levels fluctuate much less than in other wetland types.

As a result, organic matter accumulates to form thick deposits of organic soil. Most peatland plants have boreal and north-temperate distributions. Various combinations of heath shrubs, peat mosses, and sedges dominate most peatlands in the state.

SWAMPS

Swamp communities are forested wetlands with various trees, shrubs, herbs, and moss species of boreal, temperate, or coastal plain distribution. Swamps occur as wetlands in isolated depressions surrounded by upland forests, or as transitions from wetlands to uplands. Swamps are divided into two broad types, poor and rich, based on nutrient levels.

MARSHES

Marsh communities are wetlands dominated by herbaceous plants and shrubs. Trees are sparse or absent. Flood-tolerant shrubs, grasses, sedges, forbs, or aquatic plants of temperate distribution are common, with dominance depending on hydrology. Marshes occur along low-gradient streams or occupy basins where water remains near or above the surface for substantial portions of the growing season, although water levels can fluctuate significantly throughout the year. Marshes are wetter than swamps, and better drained and more nutrient-rich than peatlands. Marshes associated with sand plain basins and sandy pond shores are relatively rare and contain many coastal plain species.

RIVER CHANNELS AND FLOODPLAIN FORESTS

River channel and floodplain forest communities occupy the margins of rivers and streams. River channel communities flood frequently, and contain flood- and scour-tolerant shrubs and herbs of temperate and cosmopolitan distribution. Floodplain forests flood less frequently, typically every 1 to 2 years, and are better drained than swamps but more poorly drained than upland forests. Tall trees and a dense layer of flood-tolerant herbs of temperate or coastal plain distribution characterize floodplain forests. Non-native plants are common in both river channel and floodplain forest communities.

SEACOAST

Seacoast communities occur along New Hampshire's coast in a variety of upland and wetland settings influenced by the ocean. Estuaries occur in protected embayments influenced by tidal flow. They support several types of communities such as salt marshes, estuarine wetlands dominated by herbaceous plants. Coastal dunes form on sandy stretches of the coastline and contain a variety of shrubs and herbs. Maritime natural communities occur

along rocky shores exposed to waves and/or salt spray. In these maritime settings, marine algae occupy areas exposed at low tide, whereas herbaceous plants and shrubs occur beyond tidal reach. Most plant species in seacoast communities have coastal plain distributions.

Little Cherry Pond in the North Country region.

Ecological Regions of New Hampshire

Knowledge of New Hampshire's vegetation — combined with knowledge of climate, landforms, and soils — allows delineation of eight ecological regions. Many of these regions are easily recognized by New Hampshire's residents and visitors. They also correspond closely with ecological divisions mapped by the U.S. Forest Service.

NORTH COUNTRY

The North Country includes all of New Hampshire north of the Presidential Range. This region's terrain is a mix of mountains and large river valleys, including the upper reaches of the Connecticut, Androscoggin, and Ammonoosuc rivers. Elevations range from 700 to over 4,000 feet. Bedrock geology is complex, with a diverse mix of felsic and mafic igneous and metamorphic rocks. Metamorphic bedrock and till derived from fine-grained sedimentary rocks (pelites) are common, and weather to yield silty soils. Some of the most calcium-rich rocks in the state occur in the North Country, but they occupy only a small portion of the area. Glacial till is abundant,

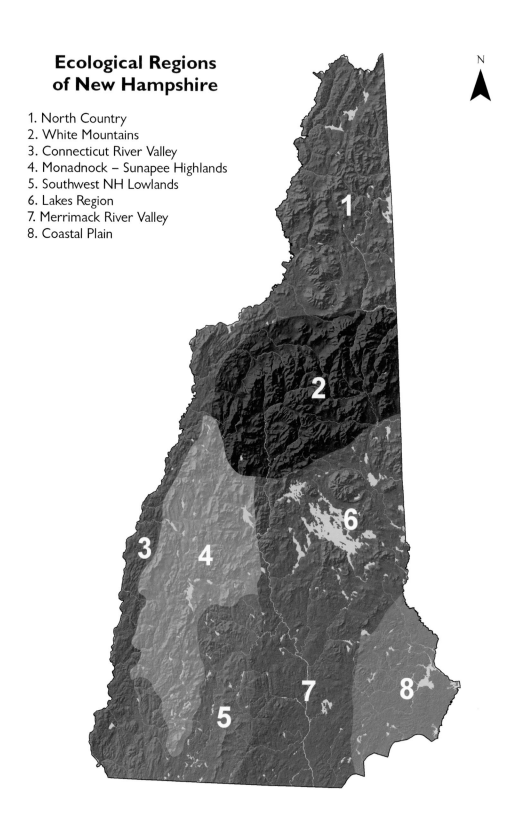

Ecological Regions
of New Hampshire

1. North Country
2. White Mountains
3. Connecticut River Valley
4. Monadnock – Sunapee Highlands
5. Southwest NH Lowlands
6. Lakes Region
7. Merrimack River Valley
8. Coastal Plain

N

Mount Franklin in the White Mountains region.

and water-deposited soil materials fill the major river valleys. Landforms include kames, eskers, and outwash and lakebed deposits. North Country communities resemble the boreal forest region of Canada. Spruce and fir forests dominate both high and low elevations, with interspersed northern hardwood forests. Peatlands, lakes, and ponds are common in the lowlands. Rich fens are occasional in areas with calcium-rich bedrock.

WHITE MOUNTAINS

The White Mountains are the dominant topographic feature of New Hampshire. The northern side of the region drains into the Connecticut and Androscoggin rivers, whereas the majority of the region drains into the Pemigewasset and Saco rivers. The upper reaches of these rivers are high-gradient, and the valleys are filled with coarse sand and gravel deposits. Regional elevations range from less than 1,000 to more than 6,000 feet. Forty-six peaks in the region exceed 4,000 feet elevation, with steep gulfs or ravines carved into the flanks of many of the higher peaks of the Presidential Range—a notable legacy of New Hampshire's glacial history. Bedrock in the region is primarily granites and schists, and surface deposits are deep glacial till overlain by acidic soils. Small areas of calcium-rich bedrock and soils occur along the western and northwestern margins of the White Mountains adjacent to the Connecticut River Valley region. The variety of plants and natural communities reflects the region's topographic relief. Alpine, sub-alpine, and rocky ridge communities are common on the bare summits of the mountains. Spruce - fir and northern hardwood forests dominate lower elevations. Lakes, ponds, and wetlands are smaller and less frequent than in other parts of the state.

CONNECTICUT RIVER VALLEY

The Connecticut River Valley is long and narrow, extending from the Massachusetts border to the southern end of the North Country. Elevations range from 300 feet in the south to more than 1,500 feet in the north. The valley contains extensive lakebed sediments from historic glacial Lake Hitchcock. Glacial till deposits occur on relatively low hills adjacent to the river, and fine, silty sediments are actively deposited along the floodplain. Most of the underlying bedrock is metamorphic, and includes most of the calcium-rich bedrock in the state. The region's lower elevations are characterized by a relatively mild climate, and as a result, numerous plants and natural communities otherwise restricted to southern parts of the state extend far up the valley. Laurentian mixed forests and Appalachian oak and pine forests dominate, and rich woods are relatively common compared to other parts of the state. Wetlands, ponds, and lakes are relatively uncommon. Rocky ridges and cliffs are occasional. Floodplain forests are common but small, as most of the high floodplains and adjacent terraces have been converted to farmland. River channel communities are nearly continuous along the river, although much of the river is regulated by dams.

Floodplain forest in Bedell Bridge State Park in the Connecticut River Valley region.

MONADNOCK-SUNAPEE HIGHLANDS

The Monadnock-Sunapee Highlands region occupies the area southwest of the White Mountains to Mount Monadnock in the southwestern part of the state. There are no major river valleys, but many tributaries to the Merrimack and Connecticut rivers form relatively steep, narrow valleys among low hills and small mountains. Ponds and lakes are common. Elevations range from about 500 feet to more than 3,000 feet on Mount Monadnock, Mount Cardigan, and Smarts Mountain. Bedrock is dominated by granite and metamorphic rocks resistant to weathering. Rocky, acidic till soils of moderate to shallow depth are common. Laurentian mixed forests are the main forest type in the region, although Acadian spruce - fir forests cap many of the higher summits. Rocky ridges, small cliffs, and talus slopes are common. Marshes, swamps, and peatlands are common, but relatively small compared to wetlands in regions with more extensive lowland areas.

Mount Kearsarge from Bog Mountain in Wilmot in the Monadnock-Sunapee Highlands region.

SOUTHWEST NEW HAMPSHIRE LOWLANDS

The Southwest New Hampshire Lowlands region occupies the relatively low elevation areas between the Connecticut River Valley and the Merrimack River Valley. The Ashuelot, Contoocook, and Piscataquog rivers are the main drainages. The terrain is complex, with rolling hills of small to moderate size, many of which are drumlins, larger bedrock-controlled hills, and narrow and broad stream and river valleys. Elevations are generally less than 1,000 feet, although some hills rise considerably higher. Bedrock is comprised mainly of granite that is resistant to weathering. Glacial till soils are mostly acidic. Laurentian mixed forests and Appalachian oak and pine forests dominate the landscape. Lakes and ponds are common, and peatlands, marshes, and swamps are abundant.

LAKES REGION

The Lakes Region has abundant small and large lakes, hills, broad plains, and several small mountain ranges including the Ossipee Mountains, the Squam Range, the Belknap Mountains, and the Blue Hills. Several moderate-sized rivers and their tributaries flow through the region, including the Saco, Bearcamp, and Winnipesaukee rivers. Elevations range from 300 to more than 3,000 feet. Glacial till covers the sloped uplands, and extensive outwash deposits, eskers, and kames occur near Ossipee. Lakes Region bedrock is primarily erosion-resistant granite and metamorphic schists. Peatlands, marshes, and swamps are numerous, including some of the largest south of the White Mountains.

MERRIMACK RIVER VALLEY

The Merrimack River Valley region occupies lowlands along the Merrimack River from Franklin to the Massachusetts border. The historic glacial Lake Merrimack once occupied much of this valley, which is now filled with abundant stratified sand and gravel overlying fine lakebed deposits. Rocky glacial till covers much of the adjacent low hilly terrain. Elevations are less than 500 feet near the river, and rise to around 1,000 feet on hilltops elsewhere. Appalachian oak and pine forests are the main forest type, with patches of historically widespread pitch pine woodlands. Wetlands are common, including peatlands, swamps, and marshes. Floodplains occur along the Merrimack River and its major tributaries. White pine is abundant throughout the region. Many temperate plants such as scarlet oak, hickories, and sassafras reach or approach the northern end of their geographic range in the Merrimack River Valley region. A number of coastal plain species are also frequent.

COASTAL PLAIN

The Coastal Plain region extends from the seacoast to 30 miles inland, encompassing the land south of the Lakes Region and east of the Merrimack River Valley. Elevations are mainly below 500 feet, although a few hills such as the Pawtuckaway Mountains exceed 1,000 feet. Metamorphic rocks underlie much of the area, and include schist and gneiss, as well as some calc-silicate rocks that contain relatively high levels of calcium. Deep surficial deposits, including sandy glacial tills and large areas of stratified sand and gravel outwash, cover most of the region. Drumlins are relatively

Rhododendron State Park in the Southwest New Hampshire Lowlands region.

West Rattlesnake Mountain and Squam Lake in the Lakes Region.

Horseshoe Pond in the Merrimack River Valley region.

Odiorne Point State Park in the Coastal Plain region.

common compared to most other parts of the state. Marine silts and clays extend up many of the river valleys within 15 miles of the coast. The Atlantic Ocean has a significant moderating effect on the climate in this part of the state. Appalachian oak and pine forests are the main forest types. Marshes and swamps are abundant throughout, and peatlands are frequent in the outwash areas. Tidal marshes, dunes, beaches, and rocky shores are unique features of the immediate seacoast vicinity. As would be expected, the largest numbers of coastal plain plant species in the state occur in the Coastal Plain region. Temperate species approaching their northern limit also occur in the region.

What This Book Does Not Cover

This book encompasses upland, freshwater wetland, and tidal wetland natural communities. It does not cover deepwater habitats of streams, rivers, lakes, ponds, or the ocean. Agricultural fields and developed lands are a fundamental part of today's New Hampshire landscape, but they are not natural communities and thus not included here.

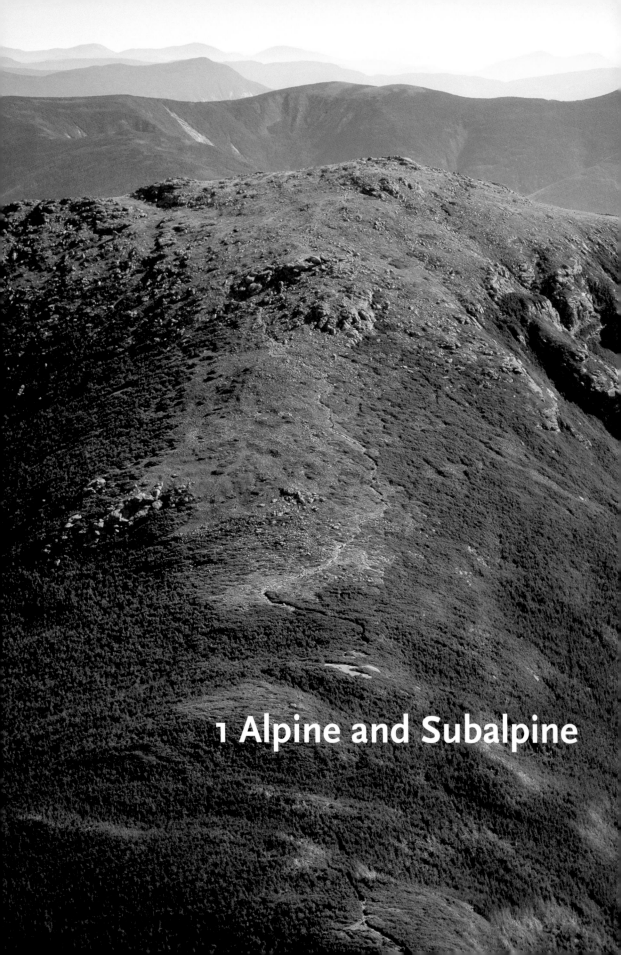

1 Alpine and Subalpine

A world of raw beauty and surprising diversity crowns New Hampshire's highest mountain peaks. In this environment, winter arrives early and stays late, and fog, rain, snow, and hurricane-force winds are a common occurrence, yet a hardy suite of alpine plants is physiologically adapted to endure these conditions. Many of the state's rarest plants also grow here, persisting in a thin zone of life between rock and sky.

Alpine and subalpine flora consists of low, mat-forming shrubs, sedges, rushes, grasses, mosses, lichens, and occasional stunted trees. Around 10,000 years ago, in the aftermath of the most recent glaciation, this vegetation covered much of New England. As the climate warmed, trees and other plants recolonized the region, relegating alpine and subalpine vegetation to higher elevations where trees could not grow. Today, New Hampshire's true alpine communities only occur on mountain summits and ridges above 4,900 feet elevation in the Presidential Range, along Franconia Ridge, and on Mount Moosilauke. Alpine vegetation is often referred to as tundra due to its similarity to Arctic tundra. New Hampshire's alpine zone is also referred to as alpine tundra. Subalpine areas, which support stunted trees in addition to alpine species, occur on exposed summits between 3,000 and 4,900 feet elevation.

The treeline marks the transition from high-elevation forest to alpine and subalpine vegetation. Its presence is largely a function of the colder temperatures, stronger and more persistent winds, late-melting snows, abbreviated

PRECEDING PAGE: Alpine and subalpine communities on the west ridge of Mount Lafayette.

Winter conditions in the alpine zone of the northern Presidential Range.

growing seasons, and nutrient-poor substrates associated with higher latitudes and
elevations. In New Hampshire, climatic
treeline occurs around 4,900 feet, though
the upper elevational limit of tree growth
varies considerably from peak to peak.

Mount Washington and other White
Mountain summits boast some of the
world's worst weather, with extremely cold
temperatures, abundant precipitation, and
one of the highest recorded wind-speeds
on Earth. Wind and blowing snow prune
tree branches so severely in this harsh environment that an upright tree cannot produce enough new growth to offset losses.
Spruce and fir trees instead adopt a twisted,
stunted life-form, often growing sideways
and low to the ground. They form waist-
high, bonsai-like patches called krummholz, a German word meaning "crooked
wood."

Alpine and subalpine plants have many
strategies for withstanding harsh physical
and climatic conditions. Nearly all are perennials with an ability to spread vegetatively, an adaptation that allows plants to

Wind-sculpted trees and
shrubs form krummholz, as
seen here on Mount Guyot.

persist without regularly reproducing from seed. Growing low to the ground
is a common strategy for conserving water and nutrients in windy conditions. Many alpine shrubs, for example, form ground-hugging cushions or
mats, and have evergreen leaves which are thick, small, and leathery, with
curled margins or dense hairs. Forbs form rosettes, whereas sedges, grasses,
and rushes form short tufts. Persistent, dead leaves shield and insulate young
shoots. Rosettes, tufts, and persistent leaves help alpine and subalpine plants
withstand physical damage and dehydration, especially in winter when vegetation is brittle and frozen soil water prevents plants from replacing the
water they lose to evapotranspiration. Evergreen leaves and an ability to store
considerable resources in root structures help some of these plants conserve
carbohydrates and nutrients, and provide a jump start on the short growing
season. Two other important adaptations to the short growing season are
the ability to photosynthesize efficiently at low temperatures and light levels,
and the production of extra anthocyanin, a red leaf pigment that enhances
uptake of solar radiation and boosts leaf temperatures.

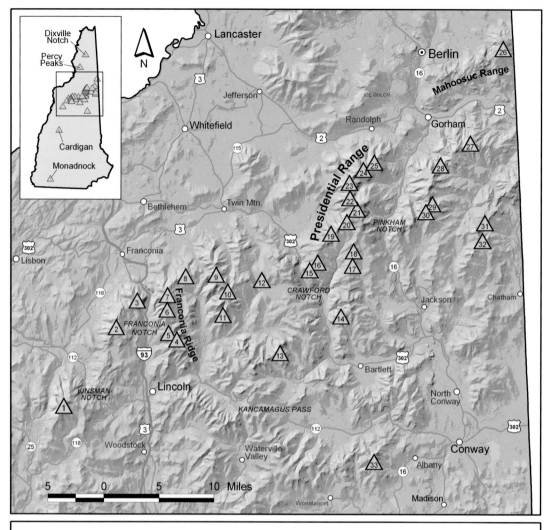

Alpine and Subalpine Areas in New Hampshire

1. Mt. Moosilauke
2. South Kinsman Mtn.
3. Cannon Mtn.
4. Mt. Flume
5. Mt. Liberty
6. Mt. Lincoln
7. Mt. Lafayette
8. Mt. Garfield
9. South Twin Mtn.
10. Mt. Guyot
11. Mt. Bond / Bondcliff
12. Whitewall Mtn.

13. Mt. Carrigain
14. Mt. Crawford
15. Mt. Webster
16. Mt. Jackson
17. Mt. Davis
18. Mt. Isolation
19. Mt. Eisenhower
20. Mt. Monroe / Mt. Franklin
21. Mt. Washington
22. Mt. Clay
23. Mt. Jefferson
24. Mt. Adams

25. Mt. Madison
26. Mt. Success
27. Shelburne-Moriah Mtn.
28. Imp Mtn.
29. Mt. Hight
30. Carter Dome
31. Eagle Crag
32. North and South Baldface Mtns.
33. Mt. Chocorua

Lapland rosebay *(left)* and alpine azalea *(right)* are low-growing alpine heath plants that have thick, waxy, evergreen leaves with curled margins, strategies that help conserve water and resources.

In terms of plant composition, New Hampshire's alpine and subalpine vegetation is more like that of the eastern Canadian arctic, hundreds of miles to the north, than the forest just hundreds of feet below. Seventy of the more than one hundred species found in the state's alpine and subalpine areas are specific to this habitat. Plant species of lower-elevation montane communities commonly occur in subalpine communities, but not in alpine communities. In addition to elevation, soil moisture and drainage, degree of exposure to wind, and length of snow-free period also affect the distribution patterns of alpine and subalpine communities and plants.

Of the seventy plant species restricted to alpine or subalpine communities, forty occur only in alpine communities. Twenty-four of the remaining thirty species occur in both alpine and subalpine communities; six of the thirty species are restricted to subalpine habitat. Due to the limited extent of the habitat, sixty-three of the seventy alpine and subalpine plant species are state threatened or endangered. Three species—dwarf cinquefoil, mountain avens, and Boott's rattlesnake root—are restricted to New England, New York, and Nova Scotia, giving alpine and subalpine areas in this part of the continent a unique floristic signature.

Alpine tundra communities on the south ridge of Mount Washington.

Alpine Tundra

New Hampshire's dramatic alpine landscapes were sculpted by ice. Continental glaciers repeatedly advanced and retreated over the past 1.8 million years, scouring the rocky peaks of the White Mountains and carving large, bowl-shaped ravines in their sides. The combination of glacial action and severe climate produced a zone of treeless alpine tundra. Today, 12 square miles of alpine tundra still occur in the Presidential Range, with another square mile on other summits. These islands of alpine tundra occur above climatic treeline, generally starting around 4,900 feet, but sometimes as low as 4,200 feet on a few exposed ridges and in snow-laden ravines.

Species diversity is uneven across the alpine zone. Lower, smaller peaks tend to be drier; they have less habitat diversity and fewer rare species. In contrast, Mount Washington's Tuckerman Ravine and Alpine Garden are home to more than half of the sixty-three rare alpine/subalpine species in New Hampshire. Three endemic or near-endemic alpine species occur in New Hampshire. Boott's rattlesnake root occurs in a few locales in the Presidential Range, New York's Adirondacks, and Vermont's Green Mountains. Mountain avens is limited to the White Mountains and two bogs in Nova Scotia. Dwarf cinquefoil is a White Mountain specialty. The dominant plants of alpine tundra are dwarf shrubs, such as bilberry, mountain cranberry, and crowberry, and tufted sedges, grasses, and rushes. Forbs account for a third of all alpine plant species, but are only abundant in wet areas such as snowbanks and large ravines.

Wind exposure, temperature, soil moisture, and drainage determine the distribution of plants and communities in the alpine zone. The degree of exposure to wind and cold is largely a function of topographic setting. Windward slopes and summits accumulate less snow than lee positions. Conversely, large ravines on east-facing slopes, such as Tuckerman Ravine, often accumulate over 50 feet of snow each winter. Snow cover remains late into spring or early summer, protecting plants underneath from extreme winter conditions. Soil moisture and drainage are a function of local topography and soil texture. A short growing season and cold climate limit mineral weathering and plant-matter decomposition. Consequently, soils are poorly developed and low in nutrients compared to soils of temperate climates. Organic matter accumulates on the surface or mixes with mineral material, forming a thin veneer over rock or mineral soil. In depressions, poorly decomposed organic soils support peatlands. Perennially wet or seepy areas under late-melting snowbanks have mucky, well-decomposed soils and support snowbank, rill stream, or ravine thicket communities. On steeper, drier terrain with coarse, well-drained soil, sedge meadows and shrub barren communities develop. Wind-exposed areas may be dominated

Mountain avens, found only in the White Mountains and a few bogs in Nova Scotia.

Boott's rattlesnake root, found only in alpine areas of the northeast.

Dwarf cinquefoil, found only in the White Mountains.

Mosaic of alpine natural communities on the summit of Mount Washington and the Alpine Garden (*lower right*). **Felsenmeer barrens** (light gray) are the most common community type here. **Bigelow's sedge meadows** (light brown) occur on upper slopes. **Sedge - rush - heath meadow** (golden brown) mixes with **diapensia shrubland** and **alpine heath snowbank** communities (maroon) on middle slopes. A patch of **black spruce - balsam fir krummholz** (dark green) dominates at lower left.

by a sparse cover of one or two species, whereas less exposed areas with more snow cover support greater plant diversity. Rocky areas such as cliffs, boulder fields, and actively shifting talus slopes have little or no soil and are sparsely vegetated.

Several unusual geologic phenomena can be observed in the alpine zone. Frequent freeze-thaw cycles cause differential movement of coarse and fine mineral material, resulting in interesting physical patterns such as rock rings, rock stripes, soil boils, and stone terraces. King Ravine, a large, steep-walled cirque on the north slope of the Presidential Range, has the remnants of a rock glacier, a moving mass of ice mixed with tons of rock. Evidence of the historic rock glacier is provided by a field of chaotic talus, dropped onto the ravine's floor by melting ice.

Summits and Upper Slopes

A dense canopy of balsam fir dominates New Hampshire's highest forests. On higher-elevation slopes, tree height diminishes and balsam fir is joined by black spruce, a true boreal tree species that grows well in harsh conditions. The low, twisted trunks and branches of these two species form **black spruce - balsam fir krummholz**. Trees here are stunted, rarely reaching more than 6 feet in height. The spruce and fir are interspersed with lesser

ALPINE NATURAL COMMUNITIES

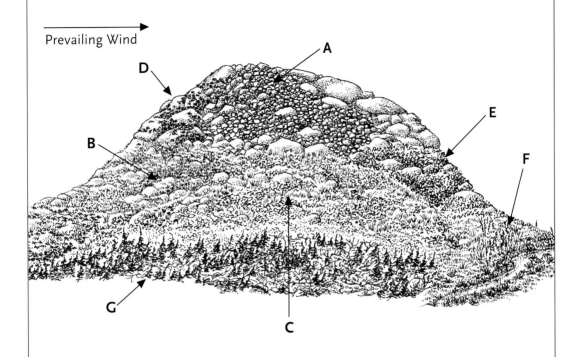

Prevailing Wind

A.
D.
E.
B.
F.
G.
C.

A. Felsenmeer barrens – *upper slope*
    Slopes of frost-fractured, lichen-covered boulders

B. Bigelow's sedge meadow – *middle left*
    High-elevation north and west slopes

C. Sedge - rush - heath meadow – *center*
    Widespread community, intermediate along environmental gradient

D. Diapensia shrubland – *upper left*
    Wind-exposed, typically convex settings

E. Alpine heath snowbank – *middle right*
    Well drained areas under late-melting snowpacks

F. Alpine herbaceous snowbank/rill – *lower right*
    Wet-mesic areas under late-melting snowpacks, along rills, and in ravines

G. Black spruce - balsam fir krummholz – *lower left and center*
    Transition zone at tree line

An idealized alpine summit in New Hampshire showing the distribution of communities relative to environmental factors, with plant scale exaggerated in relation to elevation span (~4,800 to 6,000 feet). Prevailing winds are from the west to northwest (left); maximum snow accumulation is on the lee slopes to the east (right).

Black spruce - balsam fir krummholz on Mount Adams.

Felsenmeer barren on Mount Washington.

Sedge - rush - heath meadow in the Alpine Garden on Mount Washington.

Moss-plant, found in the **alpine heath snowbank** community.

amounts of heartleaf birch, and sparse heath shrubs like mountain cranberry. This community occurs as narrow strips on slopes at treeline, as broad patches in ravines, and as isolated patches among other alpine communities above treeline. It often marks the transition between high-elevation forests and the alpine zone, though it is replaced by subalpine heath - krummholz communities at lower elevations.

Five natural communities comprise much of the alpine tundra area on open slopes and summits above the krummholz zone. On the highest peaks and ridges, jumbles of large, frost-cracked rocks form a community called **felsenmeer barrens**. Felsenmeer is a German word meaning "sea of rocks." The barrens are a product of frequent freezing and thawing in the alpine zone. Because water expands as it freezes, ice will enlarge any exposed cracks in a rock, eventually shattering it into smaller fragments. Soil only occurs in scattered patches among the rocks. As a result, vascular plants are absent or sparse, and lichens are the dominant life-form. Felsenmeer barrens are abundant in the Presidential Range, but less common in other alpine areas. Mount Washington is sometimes called the Big Rock Pile in recognition of its extensive upper slopes of felsenmeer.

Sedge - rush - heath meadow is the most common natural community in New Hampshire's alpine areas. A mix of Bigelow's sedge, highland rush, and dwarfed heath shrubs dominates most examples, with Bigelow's sedge becoming less abundant at lower elevations. It can be thought of as the principal alpine tundra community, occupying settings in the middle of the alpine zone's exposure, moisture, and elevation gradients.

Alpine heath snowbanks occur in well-drained, lee slope positions. The lower wind speeds allow snow to accumulate to a greater depth than in

sedge - rush - heath meadows. The community is characterized by the presence of Labrador tea and crowberries, as well as several rare plants such as mountain heath, moss-plant, dwarf bilberry, and alpine bearberry. **Diapensia shrublands** occupy the harshest, most windblown settings of the alpine zone, where snow does not accumulate in winter due to the strong winds. Diapensia is the dominant shrub, and grows in greater abundance than Bigelow's sedge or highland rush. Two rare plants, Lapland rosebay and alpine azalea, are most abundant in this community, though they are absent from lower-elevation examples outside of the Presidential Range. **Bigelow's sedge meadows** are dominated by Bigelow's sedge, and occur at high elevations in the Presidential Range. Other alpine species occur scattered in lower abundances. This community is most common on the upper north and west slopes of Mount Washington.

Diapensia shrubland in the Alpine Garden.

Ravines and Snowbank/Rills

The steep, wet, and snowy environments of large alpine ravines support cliff, snowbank/rill, shrub thicket, talus, and krummholz communities. Ravines feature curving, upper headwall rims of cliffs and steep rock slabs. Talus-choked landslide and avalanche gullies threaded with alpine rills streak the ravine walls. Alpine examples of **montane - subalpine acidic cliffs** (see chapter 2) often support species not found at lower elevations, and are most spectacular in the large ravines of the Presidential Range. Livelong saxifrage and nodding saxifrage are two very rare plant species restricted to alpine zone cliffs. Krummholz and shrub thicket communities often cover the floors and lower slopes of the ravines.

Bigelow's sedge on Mount Jefferson.

Bigelow's sedge meadow on Mount Washington.

Tuckerman Ravine includes snowbank/rill, shrub thicket, talus, cliff, and krummholz communities.

Alpine herbaceous snowbank/ rill in Edmands Col, with mountain avens in the foreground.

Alpine herbaceous snowbank/rill community occupies wet, lee positions under late-melting snowbanks. Most abundant on the lips, headwalls, and gullies of big ravines, this community occurs in scattered patches on higher slopes along streams and in seeps, and in the lee of moist outcrops. Late snowmelt shortens an already brief growing season, but allows alpine plants to avoid spring frosts; as a result, lowland forbs are present among the higher-elevation species. Large-leaved goldenrod, false hellebore, bluets, and bluejoint are among the species that characterize this community, which also harbors many rare, wet-site alpine plants such as hairy arnica and tea-leaved willow. Other rare plants of this community include dwarf bilberry and mountain avens.

The very rare **moist alpine herb - heath meadow** community is restricted to the Alpine Garden on the eastern shoulder of Mount Washington. It is similar to the alpine herbaceous snowbank/rill community, but lacks wetland species like hellebore, bluejoint, and bluets. The community contains numerous rarities and calciphiles (calcium-loving species). Characteristic plants include harebell, scirpus-like sedge, Boott's rattlesnake root, and viviparous knotweed.

The lower slopes of Tuckerman Ravine, Great Gulf, and other large ravines on Mount Washington support extensive **alpine ravine shrub thickets**, dominated by dense tangles of deciduous shrubs like mountain alder.

CHARACTERISTIC SPECIES OF SELECTED ALPINE NATURAL COMMUNITIES

A = Felsenmeer barren
B = Bigelow's sedge meadow
C = Sedge - rush - heath meadow

D = Diapensia shrubland
E = Alpine heath snowbank
F = Alpine herbaceous snowbank/rill

COMMON NAME	SCIENTIFIC NAME	A	B	C	D	E	F
DWARF SHRUBS							
Diapensia*	Diapensia lapponica*			o	●	o	
Alpine bilberry	Vaccinium uliginosum	o		●	o	o	
Mountain cranberry	Vaccinium vitis-idaea	o		●	o	o	
Three-toothed cinquefoil	Sibbaldiopsis tridentata			●	o		o
Bearberry willow*	Salix uva-ursi*				o		
Lapland rosebay*	Rhododendron lapponicum*				o		
Alpine azalea*	Loiseleuria procumbens*				o		
Labrador tea	Ledum groenlandicum					●	
Black crowberry	Empetrum nigrum					●	
Dwarf bilberry*	Vaccinium cespitosum*					●	●
Mountain heath*	Phyllodoce caerulea*					o	
Moss-plant*	Harrimanella hypnoides*					o	
Alpine bearberry*	Arctostaphylos alpina*					o	
Dwarf willow*	Salix herbacea*					o	o
Dwarf birch*	Betula glandulosa*					o	
Bunchberry	Cornus canadensis					o	o
Tea-leaved willow*	Salix planifolia*						o
Silver willow*	Salix argyrocarpa*						o
HERBS							
Bigelow's sedge*	Carex bigelowii*		●	●	o	o	o
Highland rush	Juncus trifidus	o	●	o	o		
Mountain sandwort	Minuartia groenlandica	o	o	o	o		
Boreal bentgrass	Agrostis mertensii		o	o			
Mountain firmoss*	Huperzia appalachiana*	o	o				
Tussock bulrush	Trichophorum cespitosum			o			o
Cutler's goldenrod*	Solidago cutleri*				o		
Starflower	Trientalis borealis					o	
Canada mayflower	Maianthemum canadense					o	
Common hairgrass	Deschampsia flexuosa					o	●
Large-leaved goldenrod	Solidago macrophylla					o	●
Bluejoint	Calamagrostis canadensis						●
Mountain avens*	Geum peckii*						●
False hellebore	Veratrum viride						●
Bluebead lily	Clintonia borealis						o
Hairy arnica*	Arnica lanceolata*						o
Hornemann's willowherb*	Epilobium hornemannii*						o
Goldthread	Coptis trifolia						o
NON-VASCULAR							
Lichens		●	o	o	o	o	
Bryophytes		o	o	o	o	o	

● = abundant to dominant   o = occasional or locally abundant   * = state threatened or endangered species

Several of the many species found in **alpine herbaceous snowbank/rill** can be seen here on the headwall of Great Gulf, including bluets, false hellebore, bluebead lily, and large-leaved goldenrod.

More than half of the rare species in the alpine zone are found in ravines and other wet snowbank communities, including hairy arnica, pictured here.

Alpine ravine shrub thickets dominated by mountain alder cover the middle and lower slopes of large ravines in the Presidential Range.

Forbs found in adjacent snowbank/rill communities are common in the understory. Deep, late-melting snowpacks bury these thickets for much of the year, and add moisture to the well-drained rocky soils. Below treeline, some talus slopes and rocky ravine-bottoms support the **subalpine cold-air talus shrubland** community with stunted spruce and alpine plants. The talus section in chapter 2 provides additional details about this community. **Labrador tea heath - krummholz**, a community characteristic of subalpine summits, occurs near treeline and in some alpine ravines.

Good Examples of Alpine Communities

In New Hampshire, alpine tundra occurs in the Presidential Range from Mount Madison in the north to Mount Pierce in the south, and on Franconia Ridge, Mount Moosilauke, Mount Guyot, and Bondcliff.

A **sheep laurel - Labrador tea heath - krummholz** community on South Baldface Mountain.

Subalpine

Several dozen subalpine peaks rise above the forest that blankets New Hampshire's White Mountains. These open, rocky exposures are below the elevation at which alpine tundra dominates. However, climatic conditions are severe enough for some alpine plants to coexist with montane species also characteristic of rocky ground communities at somewhat lower elevations. Subalpine communities are dominated by patches of spruce and fir krummholz alongside dwarf shrubs such as Labrador tea and crowberries. On many summits, barren ledges and rock outcrops exist where fires have burned, increasing soil erosion. Subalpine communities also occur near or below the widespread alpine communities in the Presidential Range and on Franconia Ridge.

Black crowberry, one of the arctic-alpine plants restricted to alpine and subalpine areas in New Hampshire.

Alpine species are less diverse and abundant in subalpine areas than they are at higher elevations. Moreover, many of the rare alpine species are scarce or absent below 4,900 feet. Thirty of the seventy alpine/subalpine species occur on subalpine peaks, but no more than fifteen occur on any one peak. Good indicators of subalpine communities include alpine bilberry, black crowberry, purple crowberry, mountain cranberry, and the lichen *cetraria islandica*. Only a few species present in subalpine areas are absent from the alpine zone, including silverling, New England northern reedgrass, Canadian mountain rice, jack pine, and northern comandra.

Mountain cranberry is found in many alpine and subalpine communities.

Environmental conditions are less extreme on subalpine summits than on alpine summits. The average temperature is higher, snow disappears more quickly, the growing season is longer, and exposure is not as severe. Nonetheless, the combination of elevation, exposure, thin soils, and deep

In New Hampshire, silverling is found only in subalpine areas, on montane cliffs, and on certain gravel barrens.

Sheep laurel - Labrador tea heath - krummholz on North Baldface Mountain.

snowpacks are still great enough to produce patches of stunted, gnarled trees among low alpine shrubs and open rock outcrops. Small-scale variations in drainage and exposure explain most of the community and species distribution patterns within the subalpine zone. Heath - krummholz communities occur on thin, well-drained soil; barrens of dwarf shrubs and highland rush occur on the most exposed areas; and bogs or dense heaths occupy poorly drained settings and lee positions that accumulate deep snow.

Heath - Krummholz

Two heath - krummholz communities occur in New Hampshire's subalpine zones. Both consist of abundant heath plants (members of the heath family), lichens, and stunted trees, and have areas of exposed outcrops, stones, and gravel. The communities typically support a cover of 40 to 80 percent knee-high heath and 20 to 80 percent krummholz. Common heath plants include Labrador tea, alpine bilberry, mountain cranberry, crowberries, and several species of blueberries. Krummholz is comprised of variable amounts of black spruce, balsam fir, and heartleaf birch trees. The krummholz layers are low and patchy, averaging less than 1.5 feet in height (with occasional taller islands of krummholz up to 5 feet in height). The shape of krummholz typically betrays the direction of prevailing winds, with branches "flagged" in one direction.

The heath - krummholz communities are distinguished from each other by elevation and species composition. Of the two, **sheep laurel - Labrador tea heath - krummholz** occurs at lower elevations, on peaks and ridges between 3,000 and 3,500 feet. In addition to sheep laurel and Labrador tea, this community also contains rhodora and mountain holly shrubs, with red

SUBALPINE NATURAL COMMUNITIES

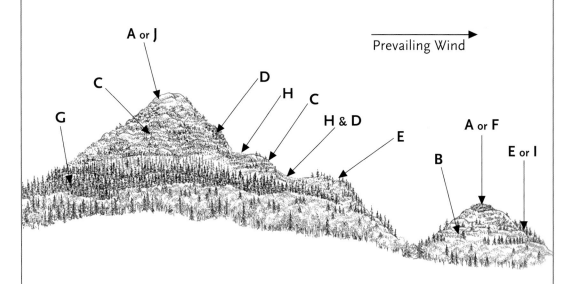

Prevailing Wind

A or J

C

G

D

H

C

H & D

E

A or F

B

E or I

A. **Subalpine dwarf shrubland** – *summits*
Wind-exposed slopes above 3,400 ft. with shallow, rocky or gravelly soil

B. **Sheep laurel - Labrador tea heath - krummholz** – *on lower peak*
Subalpine shrubland below 3,500 ft.

C. **Labrador tea heath - krummholz** – *on upper peak*
Subalpine shrubland above 3,500 ft.

D. **Wooded subalpine bog/heath snowbank** – *in lee of summits and on margins of bogs*
Concavities and slopes, often where snow accumulates

E. **Montane heath woodland** – *near forest transition*
Taller trees and heaths, moist, semi-protected, at forest transition
or on high, flat ridges

F. **Subalpine rocky bald** – *summit of lower peak*
Nearly barren, sometimes burned subalpine summits, mostly below 3,500 ft.

G. **High-elevation spruce - fir forest** – *below subalpine communities*

H. **Alpine/subalpine bog** – *perched basins*
Wet concavities and slopes above 3,000 ft.

I. **Red spruce - heath - cinquefoil rocky ridge** – *below subalpine communities*
Rock outcrop areas below 3,000 ft.

J. **Diapensia shrubland** – *summit of taller peak*
Wind-exposed settings, rarely on summits below 4,900 ft.

An idealized subalpine summit in New Hampshire showing the distribution of communities relative to environmental factors, with plant scale exaggerated in relation to elevation span (~3,000 to 4,600 feet). Prevailing winds are from the west to northwest (left); maximum snow accumulation is on the lee slopes to the east (right).

Labrador tea.

spruce occurring as part of the krummholz. Above 3,500 feet, the transition to **Labrador tea heath - krummholz** is marked by the disappearance of sheep laurel, rhodora, and mountain holly in the shrub layer, and the replacement of red spruce by black spruce in the krummholz.

In wetter, more protected areas on forested ridges below subalpine heath - krummholz communities, a dense, medium to tall shrub layer sometimes forms among moderate-height spruce and fir to create **montane heath woodlands**. Common shrubs of this community include rhodora, mountain holly, and witherod. It closely resembles the **red spruce - heath - cinquefoil rocky ridge** community in appearance (see chapter 2), but has wetter, more peaty soils, fewer rocky outcrops, and a more robust heath shrub layer.

Barrens, Balds, and Bogs

The most exposed rocky and gravelly areas on subalpine peaks above 3,400 feet support **subalpine dwarf shrublands**. Plants are a few inches high, and include crowberries, blueberries, cranberries, bilberries, and three-toothed cinquefoil. Highland rush may be present but sparse. These communities occur adjacent to heath - krummholz communities, but lack abundant sheep laurel, Labrador tea, and stunted trees. **Subalpine rocky balds** mostly occur below 3,500 feet, with a few examples found higher, and are characterized by large areas of exposed bedrock with patches of scattered subalpine vegetation relegated to cracks and hollows. Most examples of this community are associated with natural subalpine areas enlarged by fires, such as on Mount Monadnock, Mount Cardigan, and Mount Chocorua.

Subalpine dwarf shrubland on Percy Peaks, dominated by three-toothed cinquefoil.

Subalpine bogs, heath snowbanks, and heath woodlands occur in poorly drained depressions on summit ridges. Deep snowbanks in the wettest concavities facilitate the development of peat soils. These sites support two distinct communities. **Alpine/subalpine bogs** include species typically restricted to saturated conditions, such as small cranberry, baked apple berry, and hare's-tail. Steeply sloped examples of these bogs are reminiscent of sloping bogs found in more northern, maritime climates. **Wooded subalpine bog/heath snowbanks** are dominated by stunted trees, and lack the species indicative of saturated conditions in alpine/

Subalpine rocky bald on Mount Monadnock.

Alpine/subalpine bog on Mount Success. The darkest areas are mud-bottoms with the liverwort *Cladopodiella fluitans* and a moss in the genus *Drepanocladus*.

subalpine bogs. Both communities contain subalpine species such as bilberry and crowberry.

Subalpine sloping fens are steeply sloped peatlands dominated by Pickering's bluejoint, mountain avens, and peat mosses. Wetlands are unusual in steep settings, but they are occasionally made possible by a rare combination of surface runoff, seepage, cloud-intercept, cold temperatures, and low evaporation. In one example of this community, at the top edge of Cannon Cliff, portions of the peat mat periodically supersaturate and slide over the edge.

The flat to gently sloping ridgetops of the Mahoosuc, Carter-Moriah, and Baldface ranges in the White Mountains support a mosaic of wet and dry

A "heath bald" mosaic of subalpine bog and heath - krummholz communities typical of the Mahoosuc Range and Shelburne-Moriah Mountain, shown here.

Wooded subalpine bog/heath snowbank on Imp Mountain.

subalpine communities referred to as heath balds. Heath balds occur below 4,000 feet elevation, and are best developed in the Mahoosuc Range, where they extend along the ridgeline for several miles.

Good Examples of Subalpine Communities

South and North Baldface mountains, Mount Garfield, Mount Success, Mount Hight, Shelburne-Moriah Mountain, Percy Peaks, Mount Chocorua, Cardigan Mountain, and Mount Monadnock.

CHARACTERISTIC SPECIES OF SELECTED SUBALPINE NATURAL COMMUNITIES

A = Subalpine dwarf shrubland
B = Sheep laurel - labrador tea heath - krummholz
C = Labrador tea heath - krummholz
D = Wooded subalpine bog/heath snowbank
E = Montane heath woodland

COMMON NAME	SCIENTIFIC NAME	A	B	C	D	E
TREES						
Red spruce	*Picea rubens*		•			•
Black spruce	*Picea mariana*			•	•	•
Balsam fir	*Abies balsamea*		•	•	•	•
Heartleaf birch	*Betula cordifolia*		•	•		
ROBUST SHRUBS						
Mountain holly	*Nemopanthus mucronatus*		•			•
Witherod	*Viburnum nudum*					•
DWARF SHRUBS						
Labrador tea	*Ledum groenlandicum*		•	•	•	•
Sheep laurel	*Kalmia angustifolia*		•		•	•
Alpine bilberry	*Vaccinium uliginosum*	•	•	•	o	
Mountain cranberry	*Vaccinium vitis-idaea*	•	•	•	•	•
Three-toothed cinquefoil	*Sibbaldiopsis tridentata*	•	o	o		o
Black crowberry	*Empetrum nigrum*	•	o	•		
Purple crowberry	*Empetrum eamesii*	•	•	o		
Rhodora	*Rhododendron canadense*		•			•
Diapensia*	*Diapensia lapponica**	o				
Leatherleaf	*Chamaedaphne calyculata*				o	
HERBS						
Bigelow's sedge*	*Carex bigelowii**	o				
Highland rush	*Juncus trifidus*	o				
Mountain sandwort	*Minuartia groenlandica*	o				
Boreal bentgrass	*Agrostis mertensii*	o				
Tussock bulrush	*Trichophorum cespitosum*	o				
Common hairgrass	*Deschampsia flexuosa*	o				
Silverling*	*Paronychia argyrocoma**	o				
NON-VASCULAR						
Peat mosses	*Sphagnum* spp.				o	o
Other bryophytes		o	o	o	o	o
Lichens		•	•	•	o	

• = abundant to dominant   o = occasional or locally abundant   * = state threatened or endangered species

Alpine and Subalpine Wildlife

Alpine and subalpine wildlife assemblages are regionally distinctive, closely resembling those farther north or in other mountainous areas of the northeastern United States. For many species, New Hampshire's mountains are the southern limit of their global range. As with many species near their range edge, alpine and subalpine wildlife have small populations that exhibit high year-to-year variation in numbers. Overall wildlife diversity is lower in alpine and subalpine communities than in other natural community types.

Alpine

Few vertebrate species are able to breed in the alpine zone, but insects and spiders are common. Examples include small ground beetles, wolf spiders, and crane flies. Furthermore, alpine areas in the Presidential Range support robust populations of the rare, endemic White Mountain arctic and White Mountain fritillary butterflies. The White Mountain arctic is associated with communities such as **Bigelow's sedge meadow**, **sedge - rush - heath meadow**, and **alpine heath snowbank** that have an abundance of Bigelow's sedge, the larval host plant for this species. Adult White Mountain artic butterflies feed on moss campion, mountain sandwort, and mountain cranberry nectar. The White Mountain fritillary prefers wetter alpine vegetation, particularly in **alpine herbaceous snowbank/rill** communities. The larval host plant for this species is unknown, but adults gather nectar from alpine goldenrod and aster species. One notable bird species that breeds in the alpine zone is the American pipit. In New Hampshire, this bird only nests on Mount Washington.

Subalpine

Vertebrate animals are more numerous in subalpine settings than they are in the alpine zone. Krummholz vegetation offers protection from exposure and supports a relatively diverse suite of small mammals. The long-tailed shrew and rock vole are associated with rugged, rocky areas that are interspersed with krummholz, including **sheep laurel - Labrador tea heath - krummholz**, **Labrador tea heath - krummholz**, and **black spruce - balsam fir krummholz** communities. Other mammals of subalpine habitats include masked shrews, pygmy shrews, northern short-tailed shrews, and snowshoe hares. Breeding birds are also active in subalpine habitats, including the rare Bicknell's thrush and the more common blackpoll warbler. Several other species of high mountain forests also breed in the subalpine zone, including dark-eyed juncos, boreal chickadees, and gray jays.

2 Rocky Ground

Open rocky ground natural communities are scattered among New Hampshire's forested hills and mountains. Though generally small in size, these communities contribute much to the state's upland diversity. Three broad types occur: rocky ridges, cliffs, and talus. Rocky ridge is a collective term for sparsely wooded, upper slopes of ridgelines, knobs, and summits that have thin soils and frequent bedrock outcroppings. Cliffs are steep, high-angle rock outcrops. Talus slopes are jumbled blocks of angular rock that collect at cliff bases. Collectively, rocky ground communities encompass a variety of environmental conditions and support many specialized plants and animals.

Several processes are responsible for forming and maintaining New Hampshire's rocky ground communities. During the most recent ice age, glaciers scoured and gouged the landscape, exposing rocky ridges and chiseling cliffs from steep slopes. Once the glaciers receded, freeze-thaw activity and rockfall gave rise to talus slopes below cliffs. Various disturbances prevent soil accumulation and keep rocky ground communities open. Plant cover is limited by drought and fire on rocky ridges; by falling rocks, steep slopes, and severe microenvironments on cliffs; and by rockslides and rock size on talus slopes. Conversely, *lack* of disturbance allows soils to accumulate on rocky ground. In these cases, xerarch succession eventually leads to a closed canopy forest.

In New Hampshire, cliffs are most common on slopes with southern aspects. South-moving glaciers exerted pressure on the upslope (north-facing) sides of mountains and hills, producing meltwater beneath the ice that infiltrated fractures in the bedrock. Rocks embedded within the ice also abraded and smoothed the underlying substrate. Meanwhile, pressure and abrasion were lessened on the downslope (south-facing) side of the hill, resulting in lower temperatures, frost heaving within fractures, and plucking of loosened rocks. Some of New Hampshire's most prominent cliffs oppose this geographic trend, however, particularly in notches and valleys where glaciers advancing southward carved east and west facing cliffs into mountainsides.

Rocky ridges, cliffs, and talus frequently occur together, but each feature can also occur independently. For example, slopes below rocky ridges are sometimes not steep enough to include cliffs, and some forested slopes above cliffs lack rocky ridges. Bedrock porosity and degree of fracturing affect the formation of talus. At one extreme, structurally solid cliffs with few fractures produce little or no talus. Conversely, some heavily fractured cliffs in the mountains, where freeze-thaw activity is greatest, produce copious amounts of talus. Occasionally, talus accumulates to the extent that the original cliff is completely buried and obscured by its own blocky debris.

Landslides or debris avalanches are sizable slope failures that produce mosaics of outcrop, cliff, talus, and successional forest communities. Land-

PRECEDING PAGE: Open rocky ground natural communities on Cannon Mountain form a descending topographic sequence from rocky ridge, to cliff, to talus slope.

Glacial striations in exposed bedrock on Welch Mountain in Waterville Valley.

Rocky ground communities often occur together in a descending topographic sequence like this one at The Ledges near Newfound Lake in Alexandria.

slide scars are visible as linear tracks on mountain slopes where rock, soil, and vegetation slumped and slid down during intense rain events. Landslides obliterate the existing forest, creating opportunities for pioneer and early successional species. The White Mountains currently have more than 500 landslide scars in various stages of recovery. The upper parts of landslide tracks consist of cliff, outcrop, and talus material. The process of forest recovery may take centuries here. Lower portions of landslide tracks are characterized by substrates of basal till and deposits of mixed rubble debris,

ROCKY GROUND NATURAL COMMUNITIES (NORTHERN NEW HAMPSHIRE)

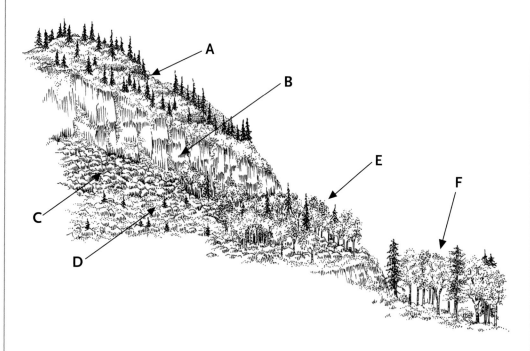

A. **Red spruce - heath - cinquefoil rocky ridge**
Upper slopes, ridges, and summits with many bedrock outcrops and northern conifers

B. **Montane - subalpine acidic cliff**
Steep, sparsely-vegetated outcrops with lichens, mosses, and microorganisms

C. **Montane lichen talus barren**
Scattered vascular plants among large, lichen-covered rocks

D. **Subalpine cold-air talus shrubland**
Stunted spruce and dwarf alpine heath shrubs on large boulders, with late-melting ice beneath boulders

E. **Birch - mountain maple wooded talus**
Talus slopes with birches, northern conifers, shrubs, vines, and herbs

F. **Northern hardwood - spruce - fir forest**
Forest communities often occur on stabilized talus slopes

An idealized rocky ground natural community sequence in northern New Hampshire. Other community combinations are possible.

ROCKY GROUND
NATURAL COMMUNITIES
(SOUTHERN NEW HAMPSHIRE)

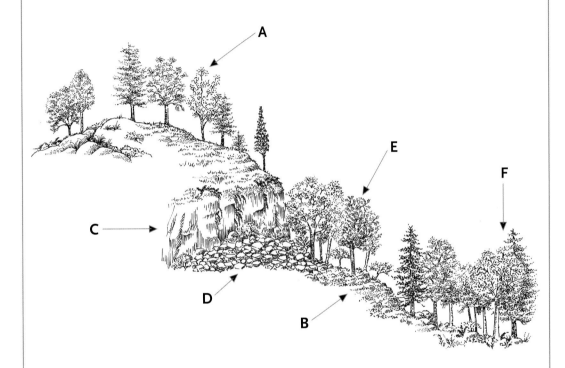

A. **Appalachian oak - pine rocky ridge**
Upper slopes, ridges, and summits with many bedrock outcrops and oaks

B. **Red oak - ironwood - Pennsylvania sedge woodland**
Open, lawn-like understories on mid to upper slopes

C. **Temperate acidic cliff**
Steep, sparsely-vegetated rock outcrops

D. **Temperate lichen talus barren**
Scattered vascular plants among lichen-covered rocks

E. **Red oak - black birch wooded talus**
Talus slopes with oaks, birches, shrubs, vines, and herbs

F. **Hemlock - beech - oak - pine forest**
Forest communities on and below stabilized talus slopes

An idealized rocky ground natural community sequence in southern New Hampshire. Other community combinations are possible.

Landslide scars occur on steep slopes and can contain mosaics of rocky ground communities, as seen here in the White Mountains.

sometimes appearing plastered to the exposed bedrock. Lower tracks succeed to forest more rapidly than the open rock areas above.

Rocky ground moisture, temperature, and nutrient levels vary with slope steepness, aspect, elevation, latitude, and rock and soil characteristics. Steep, south-facing slopes at low elevation in southern New Hampshire are warm and dry, whereas steep, high-elevation, north-facing slopes in northern New Hampshire are moister and colder. In addition, cold, wet microsites occur on cliffs where water seeps through fractures in the bedrock, in deep, boulder-choked gorges, and at the base of large talus slopes. The latter are influenced by late-melting ice and cold air settling among the boulders.

Most rocky ground settings in New Hampshire have thin, acidic soils that are low in nutrients and have a limited capacity to retain moisture. Water drains rapidly from steep, convex, thin-soiled ridges and cliffs, and disappears below large talus boulders. Soils with high nutrient levels are rare in the state's rocky ground communities, although enriched zones can develop on talus slopes at the bases of cliffs where organic matter and nutrients accumulate. In addition, some combinations of rock type, fracturing, weathering, and seepage deliver higher levels of calcium and other mineral nutrients to the rooting zones of plants.

Rocky ground plants must be efficient at acquiring and using water and nutrients. One adaptive strategy is an emphasis on root growth at the expense of above-ground stem or branch growth. This enables plants to acquire and conserve as much water as possible, reduce evaporative losses, and live within the limits of the nutrient-poor soils. Stunted or gnarled trees are common on ridges, cliffs, and talus, a reflection of the limited resources available for above-ground growth, and the pruning of branches in settings exposed to strong wind or falling rocks. In addition to the stress of limited water and nutrients, rocky ridge settings are also prone to fire, and many species are adapted to survive or benefit from periodic burning.

New Hampshire's rocky habitats support many common plants, 75 rare vascular plants, and many rare mosses and liverworts. Most of the rare plants have centers of distribution farther north or south than New Hampshire, reflecting the state's transitional location between temperate and boreal climate regions. Cold microsites on north-facing cliffs, or in cold, ice-filled talus gorges, harbor some of the southernmost low-elevation populations of alpine or boreal climate species. Warm, south-facing slopes, in contrast, harbor the northernmost populations of some temperate climate species. For example, red oak and eastern red cedar are temperate species that reach elevations of 2,200 and 1,500 feet, respectively, on warm, south-facing ridges and cliffs as far north as the White Mountains.

Pitch pines and lichens survive in the dry conditions of a rocky ridge community.

Rocky Ridges

New Hampshire's many open rocky summits, knobs, and ridgelines offer sweeping views of the landscape. These rocky ridges formed where glaciers scoured hillsides down to bedrock, and scant soils and periodic fires have hindered revegetation. Rocky ridge natural communities feature discontinuous canopies of short, gnarled trees, and open understories with patches of low-growing blueberries, huckleberries, grasses, and lichens on thin, dry soils.

Rocky ridges are the domain of oaks, pines, and spruce, wide-ranging species that compete effectively under stressful conditions. Oaks dominate on ridges below 1,700 feet elevation in central and southern New Hampshire, with lesser amounts of white or pitch pine present. Red spruce, red pine, and jack pine dominate colder-climate rocky ridges at higher elevations, mostly in central and northern parts of the state. In all of these settings, the understories are characterized by dwarf shrubs, composites (asters and goldenrods), graminoids (grasses, sedges, and rushes), and lichens. Twenty-six rare plants occur in New Hampshire's rocky ridge communities, half of which are composites or graminoids.

Dry, nutrient-poor soils and fire are the primary factors affecting rocky ridge vegetation. Thin soils have an inherently low capacity to store nutrients, and they dry readily on the convex upper slopes, leading to frequent and extended droughts. These conditions favor the growth of oaks, pines, and other plants that can tolerate harsh conditions. Oaks and pines produce leaf litter that is high in lignin and low in nitrogen, characteristics that slow decomposition rates. Organic matter accumulates and decomposes slowly,

View of Mount Cardigan from a rocky ridge on Orange Mountain.

limiting nutrient availability. Plant growth is slow, and there is little above-ground biomass. Rocky ridge plants can benefit from a short-term flush of nutrients following fires, but nutrients are quickly lost to leaching. Rocky ridges comprised of bedrock types that bear higher levels of calcium sometimes support rich-site plants, but the calcium difference is not enough to modify the overall biomass production or change the dominant plants.

Periodic fires on rocky ridges perpetuate site conditions and vegetative composition. Positioned high in the landscape, rocky ridges are subject to frequent lightning strikes. Leaves and twigs of pine, scrub oak, and other species contain volatile compounds that render the dry duff layer particularly flammable. Once a fire is started, it spreads quickly across the steep slopes, burning dense patches of blueberries and huckleberries. Severe fires can burn all the above-ground vegetation, as well as roots and other organic matter in the soil. Consequently, fires can reduce the total volume of soil, leave mineral material more susceptible to erosion, and expand the extent of barren outcrops. Once soil is removed, it can take centuries to redevelop. Most New Hampshire rocky ridge communities have experienced fires of natural or human origin. Prior to fire-suppression efforts, human-caused fires enlarged open rocky ridges in many parts of the state.

Many plants of rocky ridge communities are adapted to drought conditions and fire. For example, plants on rocky ridges generally allocate more resources to root growth than plants of moist habitats, facilitating the uptake of limited water and soil resources. Above-ground adaptations to fire include re-sprouting from roots or root crowns, a phenomenon illustrated by blueberry, huckleberry, immature oak, pitch pine, and red pine. Mature pine and oak trees develop thick, insulating bark. Jack pines release seeds from cones that are heated by fire, and germinate best in exposed mineral soils. Without periodic fire, fire-dependent species would diminish and be replaced by fire-intolerant species such as birch, maple, and white pine. The return intervals necessary to perpetuate fire-dependent vegetation range from 50 to 200 or more years, depending on the species. One can discern the timing of the most recent fire by the abundance of fire-dependent trees of the same age or, conversely, by the oldest age of fire-intolerant species.

The distribution of rocky ridge trees reflects differences in regional fire frequency and individual fire tolerances. Dry oak forests and woodlands of temperate regions and coniferous forests of boreal regions are both maintained by periodic fire. For example, in the southeast boreal region of Canada and the Great Lakes states, large fires support fire-tolerant red and jack pine stands and limit fire-intolerant red spruce. By contrast, the north temperate (Laurentian) forests that dominate much of New Hampshire burn less frequently. Having only smaller, less frequent fires leads to an increased cover of spruce and more scattered stands of red or jack pine.

Fern-leaved false foxglove. In New Hampshire, this plant is mostly restricted to openings within oak-dominated rocky ridge communities.

Smooth sandwort, a rare plant of open rocky ridges in central New Hampshire.

Elevation directly affects the distribution of rocky ridge trees in New Hampshire, though its effect is modified somewhat by other site conditions. In general, oaks dominate below 1,700 feet in elevation, red spruce dominates above 2,500 feet, and subalpine vegetation appears above 3,500 feet. A climatic transition from temperate to montane boreal between 1,700 and 2,500 feet in elevation renders species composition less predictable. Almost any combination of red oak, red spruce, red pine, white pine, or jack pine can occur here, and there may be no clear dominant. Species composition in this elevation range is a function of site conditions and site history, such as which species were first established, local seed sources, elevation, forest management, and fire history. Frequent fire favors red oak, red pine, pitch pine, or jack pine. Conversely, an absence of fire will favor the growth of red spruce.

A white oak tree in an **Appalachian oak - pine rocky ridge** on Wantastiquet Mountain in Hinsdale.

Southern and Central New Hampshire

Low-elevation (below 1,000 feet) rocky ridges in the southern third of the state support Appalachian oak species, including white, black, chestnut, and scrub oaks, pitch pine, and sometimes shagbark hickory. Three distinct woodland natural communities occur on rocky ridges at low elevation. **Appalachian oak - pine rocky ridges** are the most widespread, characterized by a mix of red and other oaks, occasional pines, and in some instances dense thickets of scrub oak. Common shrubs include lowbush blueberry, black huckleberry, hillside blueberry, maple-leaved viburnum, creeping juniper, and sweet fern. Chestnut oak is the principal tree of **chestnut oak forest/woodlands**. This species is a dry-site specialist that reaches the northeast end of its range in New Hampshire, and it has a more limited distribution in the state than other Appalachian oaks. **Pitch pine rocky ridges** are rare in the state. Pitch pine is a fire-adapted tree restricted to the Appalachian

Pitch pine rocky ridge on Moose Mountain in Brookfield.

Red oak - pine rocky ridge at Ellis Hatch Wildlife Management Area in Brookfield.

Red oak - ironwood - Pennsylvania sedge woodland on Middle Mountain in Pawtuckaway State Park.

Circumneutral rocky ridge on Holts Ledge in Lyme.

Mountains and the northeastern coastal plain. It reaches its northern limit in the Merrimack Valley, Ossipee region, and adjacent parts of Maine.

In southern and central New Hampshire, most oaks become scarce as elevation increases and, with the exception of red oak, disappear above 1,000 feet. The **red oak - pine rocky ridge** community is the most common rocky ridge type between 1,000 and 2,000 feet elevation. The community occurs as high as 2,200 feet elevation on warm, south-facing slopes of the Saco River valley in the White Mountains. Red spruce and red pine increase above 1,700 feet, and may co-dominate with red oak in some areas. Red oak is a broadly adapted temperate species, most abundant on dry sites where trees of mesic sites cannot survive.

The preceding four rocky ridge communities have southern plant species that decline with increasing elevation, reaching their maximum elevation and disappearing somewhere between 1,000 and 1,700 feet. These plants include black huckleberry, little bluestem, sassafras, eastern red cedar, ground juniper, bastard toadflax, pale corydalis, mountain laurel, and scrub oak. Fern-leaved false foxglove is a rare plant limited to oak-dominated rocky ridges in New Hampshire.

Two uncommon rocky ridge communities occur on slopes with elevated calcium levels. The more frequent of the two, **red oak - ironwood - Pennsylvania sedge woodland**, features an open understory dominated by grasslike lawns of Pennsylvania sedge. This community can also be distinguished from other rocky ridge communities by the abundance of ironwood in the sub-canopy, and herbs such as acute-lobed hepatica, rusty woodsia, and rock sandwort. It often occurs on slopes above **rich red oak rocky woods** or **rich Appalachian oak rocky woods** (see the rich woods section in chapter 3). **Circumneutral rocky ridge** is a very rare community in New Hampshire,

occurring on sloping, calcareous bedrock outcrops at the tops of two cliffs in the western part of the state. The community supports rare snowy aster and creeping juniper, two plant species with different continental ranges that converge in New England. The range of snowy aster is centered on calcium-rich prairies in the Midwest and south, whereas creeping juniper is a more northern, transcontinental plant of rocky habitats.

Central and Northern New Hampshire

Tree composition shifts from oak to northern conifer dominance at around 2,000 feet elevation. Four coniferous rocky ridge natural communities occur between 2,000 and 3,500 feet, distinguished by different combinations of red spruce, red pine, and jack pine. The tallest trees in each community are generally 10 to 40 feet tall, but they can be taller under favorable conditions. Northern herbs, shrubs, and other trees, including Rand's goldenrod, three-toothed cinquefoil, and mountain ash, accompany the transition from oak and other temperate species to northern conifers. Rare species include silverling, smooth sandwort, rock sandwort, Canadian mountain rice, and Douglas' knotweed. On exposed settings above 3,500 feet in elevation, dwarf trees and alpine plants replace northern herbs, shrubs, and trees, indicating the transition to subalpine communities (see chapter 1).

Red spruce - heath - cinquefoil rocky ridge on South Baldface Mountain in Chatham.

Red spruce - heath - cinquefoil rocky ridge is the most common of the four rocky ridge communities in central and northern New Hampshire. The community occurs between 1,700 and 3,000 feet elevation. The dominant tree, red spruce, is restricted to the northern Appalachian region, in contrast to the more boreal distributions of both jack and red pine. Red spruce is fire-intolerant, but readily colonizes the thin soils of rocky ridges where it mixes with red oak (at elevations below 2,000 feet) and red pine (up to about 2,700 feet). In the absence of fire, soils accumulate and red spruce rocky ridge communities can succeed from woodlands to forests.

Jack pine rocky ridge on Welch Mountain in Waterville Valley.

Jack pine rocky ridge communities occur at a few sites in the White Mountains between elevations of 1,800 and 3,900 feet. Jack pine is a fire-dependent species of boreal regions in eastern North America. In New Hampshire, it is rare and at the southeastern limit of its range. Most jack pine cones are serotinous, meaning they open late, requiring heat to release their seed. Non-serotinous cones are also present, however, and may be important for the perpetuation of jack pine stands that burn infrequently.

Red pine rocky ridge communities occur at elevations of 1,400 to 2,700 feet in northern, mountainous parts of the state. Red pine is a boreal species that reaches its southeastern limit in New Hampshire, but it is not rare. It occurs as a scattered tree farther south and at lower elevations, but only occurs as large stands in the mountains. Red pine bark thickens with age, and

Red pine rocky ridge on South Moat Mountain in Albany.

by 70 years is thick enough to protect the tree from fire. As a result, red pine is favored by fire-return intervals of 70 or more years. In the absence of fire, red spruce may replace red pine as the dominant tree in higher elevation examples, shifting the community type to red spruce - heath - cinquefoil rocky ridge.

Montane heath woodlands occur on moist, high-elevation ridgetop settings near or at the transition to the subalpine zone. They most closely resemble the red spruce - heath - cinquefoil rocky ridge community in appearance. A shallow and moist organic soil layer, fewer rocky outcrops, and a more robust heath shrub layer distinguish montane heath woodlands from other montane rocky ridge types. Characteristic species include balsam fir, red and black spruce, rhodora, mountain holly, sheep laurel, witherod, and peat mosses.

Montane heath woodland on Zealand Ridge in the White Mountains.

Good Examples of Rocky Ridge Natural Communities

SOUTHERN AND CENTRAL NEW HAMPSHIRE

Appalachian oak - pine rocky ridge: Rocky Ridge and South Mountain in Pawtuckaway State Park (Nottingham), Mount Wantastiquet (Hinsdale), and Rattlesnake Mountain (Rumney).

Chestnut oak forest/woodland: North Mountain in Pawtuckaway State Park (Nottingham) and Dumplingtown Hill (Raymond).

Pitch pine rocky ridge: Rock Rimmon Park (Manchester) and in the Moose Mountains (Brookfield).

Red oak - pine rocky ridge: Mount Stanton (Bartlett), White Ledge (Albany), Crotched Mountain (Francestown), and Moat Mountain (Conway).

Red oak - ironwood - Pennsylvania sedge woodland: West Rattlesnake Mountain (Holderness), Daniels Mountain (Hinsdale), Warwick Preserve (Westmoreland), and Pawtuckaway State Park (Nottingham).

Circumneutral rocky ridge: Holts Ledge (Lyme).

CENTRAL AND NORTHERN NEW HAMPSHIRE

Red spruce - heath - cinquefoil rocky ridge: Good examples occur on Mount Monadnock (Jaffrey), South Baldface Mountain (Chatham), the Squam Range (Campton), and Percy Peaks (Stark).

Jack pine rocky ridge: Carter Ledge on Mount Chocorua (Albany) and Welch Mountain (Thornton/Waterville Valley).

Red pine rocky ridge: Iron Mountain (Bartlett), Owls Head (Benton), Peaked Mountain (Conway), and Black Mountain (Haverhill).

Montane heath woodland: Zealand Mountain (Lincoln), Whitewall Mountain (Lincoln), and at the transition to the subalpine communities on Mount Chocorua (Albany) and South Baldface (Chatham).

CHARACTERISTIC PLANTS OF SELECTED ROCKY RIDGE NATURAL COMMUNITIES

A = Red spruce - heath - cinquefoil rocky ridge
B = Red pine rocky ridge
C = Red oak - pine rocky ridge
D = Appalachian oak - pine rocky ridge
E = Red oak - ironwood - Pennsylvania sedge woodland

COMMON NAME	SCIENTIFIC NAME	A	B	C	D	E
TREES						
Red spruce	Picea rubens	●	o	o		
Paper birch	Betula papyrifera	o	o			
Red pine	Pinus resinosa	o	●	●	o	
White pine	Pinus strobus	o	o	●	●	
Red oak	Quercus rubra	o	o	●	●	●
White oak	Quercus alba				●	
Black oak	Quercus velutina				●	
Eastern red cedar	Juniperus virginiana				o	
Ironwood	Ostrya virginiana				o	●
Shagbark hickory	Carya ovata					●
White ash	Fraxinus americana					●
Sugar maple	Acer saccharum					●
SHRUBS						
Rhodora	Rhododendron canadense	o				
Witherod	Viburnum nudum	o	o			
Mountain holly	Nemopanthus mucronatus	o	o			
Mountain ashes	Sorbus spp.	o	o			
Sheep laurel	Kalmia angustifolia	●	●			
Bearberry	Arctostaphylos uva-ursi		o	o		
Three-toothed cinquefoil	Sibbaldiopsis tridentata	●	o	o		
Lowbush blueberry	Vaccinium angustifolium	●	●	●	●	
Velvet-leaf blueberry	Vaccinium myrtilloides	o	●	●	●	
Black huckleberry	Gaylussacia baccata		●	●	●	
Sweet fern	Comptonia peregrina			●	●	
Ground juniper	Juniperus communis			o	o	
Scrub oak	Quercus ilicifolia				●	
HERBS						
Silverling*	Paronychia argyrocoma*	o				
Rand's goldenrod	Solidago simplex ssp. randii	●	o			
Douglas' knotweed*	Polygonum douglasii*		o	o		
Pale corydalis	Corydalis sempervirens		o	o	o	
Fern-leaved false foxglove*	Aureolaria pedicularia var. int.*			o	o	
Poverty oatgrass	Danthonia spicata	o	o	o	o	
Bristly sarsaparilla	Aralia hispida	o	o	o	o	
Common hairgrass	Deschampsia flexuosa	●	●	●	●	●
Pennsylvania sedge	Carex pensylvanica			o	o	●
Wild columbine	Aquilegia canadensis			o	o	o
Pussytoes	Antennaria spp.			o	o	●
Sweet goldenrod*	Solidago odora*				o	
Wood anemone	Anemone quinquefolia					o
Blunt-lobed hepatica	Anemone americana					o
Rock sandwort*	Minuartia stricta*					o
Rusty woodsia	Woodsia ilvensis					o
Blunt-lobed woodsia*	Woodsia obtusa*					o
NON-VASCULAR						
Mosses & Lichens		●	●	●	●	o

● = abundant to dominant   o = occasional or locally abundant   * = state threatened or endangered species

Whitehorse Ledge in Hale's Location supports a **temperate acidic cliff** community.

Cliffs

Cliffs are among New Hampshire's most distinctive and recognizable natural features. Though they may appear barren from a distance, up close cliffs support a diversity of life. Species with the ability to survive in a severe and challenging environment grow directly on bare rock, anchored in cracks, or perched on narrow ledges. The most conspicuous life-forms on cliffs are small rock-hugging lichens and bryophytes, along with occasional vascular plants. Microorganisms—including fungi, algae, and cyanobacteria—are also common, often appearing as dark stains and streaks on the rock.

New Hampshire cliffs harbor many rare bryophytes and more than thirty rare vascular plant species, including species with arctic-alpine, montane, and temperate distributions. Some of the species are relicts: remnants of larger, more widespread populations that were present during colder or warmer climatic periods. Relict plant populations persist in cold microclimates on cliffs in mountain notches and ravines. Such areas, called refugia, serve as important sources of genetic variation and propagules for the establishment of new populations under shifting climate conditions.

Most vascular plant species on cliffs also occur in other rocky or open habitats—relatively few are entirely restricted to cliffs. Perennial ferns, grasses, sedges, and composites comprise half of all vascular plants on cliffs, and nearly two-thirds of all herbaceous species. Woodsias, wood ferns, polypodies, cliffbrakes, spleenworts, and fragile ferns are among the fern species that may be present. Heaths and other dwarf shrubs are also common, particularly on dry or exposed high-elevation cliffs. Trees are sparse and small for their age. Vascular plants grow from crevices in the cliff, on organic mats established by smaller life-forms, and on horizontal ledges and other low-angle areas with accumulated soil. New Hampshire's cliffs support more than 170 species of vascular plants. The total species diversity of bryophytes

Cannon Cliff is the largest cliff in New England.

Composites, grasses, sedges, and shrubs are common types of plants on cliffs. Seen here are scirpus-like sedge, New England northern reed-grass, mountain alder, and the yellow flowers of shrubby cinquefoil and Rand's goldenrod.

Mosses and liverworts are common on cliffs and are particularly abundant around moist crevices.

Livelong saxifrage, a plant of calcareous cliffs, exudes calcium along its leaf edges.

is unknown, but on well-studied cliffs, bryophytes and vascular plants are present in approximately equal numbers.

Many cliffs have microsites, small areas with different combinations of moisture, temperature, nutrients, rock type, fracturing aspect, and exposure to sun and wind. Species composition and diversity on cliffs is more closely related to microsite diversity than to cliff size, and can vary considerably from one cliff to another. Differences in species composition associated with climate and nutrients are the primary factors used to classify cliff natural communities.

Climate on New Hampshire cliffs varies with elevation and latitude. The colder climate on cliffs at higher elevations and latitudes results in greater physical weathering. For example, fracturing and erosion of rock from water freezing in cracks is most pronounced on alpine and montane cliffs. Conversely, cliffs in warmer, wetter climates experience more rapid chemical weathering. On a single cliff, climate also varies on a local scale with wind exposure, aspect, and shading. At a given cliff, climate conditions may be harsher or milder than those of adjacent level ground. Greater wind exposure and lack of insulating snow render conditions more severe. The angle of the sun's rays is also important. On south-facing cliffs, a high summer sun and a low winter sun reduce temperature and moisture stress. North-facing cliffs receive no winter sun and minimal summer sun, and therefore are colder simply by being in the shade most of the time.

Moisture availability varies greatly among and within cliffs, producing corresponding shifts in vegetation. A cliff receives moisture from four possible sources: precipitation, surface runoff, condensation, and seepage from fractures. The small watershed and steep angle of cliffs limit water inputs from precipitation and surface runoff. High-angle slabs (cliffs with few fractures) are often dry, and support plants and lichens adapted to xeric conditions. In other areas, thin surface soils on ledges or slabs intercept and store sufficient moisture to support vegetation common to rocky ridge habitats, including trees and heath shrubs. Vegetation can be very different where water seeps from fractures in a cliff. Perennial or nearly perennial seeps can support plants typical of fens and swamps, such as round-leaved sundew and violets. Condensation on the relatively cool thermal mass of the cliff during warm, humid periods may be an important source of water for non-vascular life-forms.

Cliff vegetation also responds to nutrient availability. Sparse soil and correspondingly low nutrient levels lead to low rates of biomass production, among the lowest of any habitat type in the world. Where soils do exist, the nutrient concentration ranges from very low to very high. The most pronounced differences in vegetative composition occur between areas with differing calcium concentrations and pH levels. Calcium levels in soil water

Seasonal and perennial seepage from fractures and runoff from above are important sources of water and nutrients for some organisms. The cliffs of Devil's Hopyard in Stark contain extensive seepage zones and an abundance of mosses and liverworts.

can be 20 to 100 times higher under circumneutral conditions (pH of 6 to 7) than under acidic conditions (pH of 4 to 5). Such variation sometimes occurs across a single cliff. Acidic, nutrient-poor microsites are common in areas that receive only surface runoff and precipitation. Circumneutral, nutrient-rich conditions are often associated with fractures fed by seasonal groundwater seepage, particularly in overhanging areas protected from surface runoff and precipitation. Calcium differences have little apparent effect on productivity and biomass, but can dramatically affect species composition. Calciphiles—plants that thrive in calcium-rich soils—are characteristic of cliff microsites with high levels of calcium and a circumneutral pH. Cliffs with calcium-rich microsites harbor a diverse assemblage of calciphilic mosses, liverworts, and vascular plants not found on acidic cliffs. Many calciphilic species are rare.

Levels of calcium and other nutrients on cliffs are determined by the type of bedrock, amount of fracturing, and flow of water through fractures to the rooting zone. Readily weathered, calcium-rich bedrock produces circumneutral conditions, which enhance soil development and microbial activity, increase the levels of plant nutrients like phosphorus and molybdenum, and decrease harmful levels of soluble manganese and aluminum. Most of New Hampshire's bedrock is comprised of silica-rich and calcium-poor granite and schist, which have extremely slow weathering rates, yield acidic soils with low levels of calcium, and are less conducive to plant growth. The degree of bedrock fracturing also affects the potential yield of calcium and other nutrients from the rock, by determining the surface area exposed to chemical and physical weathering. Fractures enable the flow of water through the rock, transporting mineral nutrients to seepage areas on the cliff face. Each of these interrelated factors can contribute to significant changes in nutrient level over a short distance, often with corresponding shifts in species composition.

Cliffs afford protection from losses to fire, herbivores, and humans. Thus, adaptations common among plants in more productive habitats, such as rapid growth and disturbance tolerance, do not confer an advantage on cliffs. Instead, natural selection favors a plant's ability to survive in highly specialized conditions, such as those of dry, acidic sites or cold, wet, calcium-rich sites. For example, the ability of ferns to use water more efficiently than most flowering plants and conifers may explain their prevalence on cliffs. In general, cliff plants are adapted to small living spaces, limited resources, and particular microsite conditions. They require few resources and grow slowly. Desiccation-tolerant conifers and heath shrubs are common on dry microsites such as cliff brows or thin-soiled ledges. Bryophytes and lichens are also particularly well adapted to endure drought and other extreme cliff conditions. These non-vascular organisms limit physiological activity during periods of stress, and respond rapidly to favorable conditions. Consequently, bryophytes and lichens can often thrive where vascular plants cannot.

There are two broad groups of cliff communities in New Hampshire: temperate and montane - subalpine. Individual examples of each community exhibit substantial variation in size and microsite complexity, including factors such as fracturing, soil availability, and moisture conditions. These factors in turn lead to differences in species richness and composition. Some species, such as common hairgrass, lowbush blueberry, bush honeysuckle, meadowsweet, and long beech fern, occur on both temperate cliffs and montane - subalpine cliffs.

Temperate Cliffs

Temperate cliffs support species of north-temperate climates, including plants with eastern deciduous forest, Alleghenian, and southern boreal distributions. Most temperate cliffs are less than 150 feet in height, and are partially or entirely shaded.

Temperate acidic cliffs are the most common type of cliff in New Hampshire, and occur throughout the state below 2,200 feet elevation. Common herbaceous plants include common hairgrass, rock polypody, hay-scented fern, marginal wood fern, fragile fern, wild columbine, and a wide variety of asters, goldenrods, grasses, and sedges. Pale corydalis, fern-leaved false foxglove, and lowbush blueberry occur on dry microsites. Bush honeysuckle is a common woody plant, although it is not always abundant. Oaks, white pine, pitch pine, eastern red cedar, and ground juniper occur on more southern or lower-elevation cliffs, and occasionally on south-facing cliffs in the White Mountains. Red spruce, red pine, yellow birch, and paper birch are more

Temperate acidic cliff at Cape Horn in Northumberland.

common on northern or higher-elevation cliffs. Wet seepage areas support abundant mosses and liverworts, and vascular plants typical of fens and swamps such as round-leaved sundew, bluets, violets, meadowsweet, dwarf raspberry, whorled aster, golden saxifrage, and bluejoint. The gorges at The Flume in Lincoln and Devil's Hopyard in Stark are among New Hampshire's most extensive cliff seeps.

Temperate circumneutral cliffs are less common than their acidic counterparts. They share many of the same species, but differ by the presence of plants indicative of elevated calcium and pH conditions. These indicator plants include fragrant fern, slender cliffbrake, maidenhair spleenwort, smooth woodsia, early saxifrage, northern white cedar, and red elderberry. The most common species on temperate circumneutral cliffs are rusty woodsia and harebell, plants that can also tolerate moderately acidic sites. The rare fragrant fern, a species restricted to overhangs on circumneutral cliffs, is a reliable indicator of high levels of calcium. It grows only where there is seasonal circumneutral seepage. Calciphilic bryophytes also occur near cracks that emit calcium-rich seepage water. Circumneutral cliff communities may occupy most of the area of a cliff, or be restricted to specific zones within a larger acidic cliff.

Temperate circumneutral cliff at Diamond Peaks north of Umbagog Lake.

Montane - Subalpine Cliffs

Montane - subalpine cliffs support plant species with north-temperate, boreal, and alpine distributions. Most montane - subalpine cliffs are large, exposed to wind, and above 2,200 feet in elevation. Some of them exceed 150 feet in height.

Montane - subalpine acidic cliffs occur primarily in the White Mountain region. Among cliffs, montane and alpine species largely or entirely restricted to this community include heartleaf birch, mountain cranberry, three-toothed cinquefoil, velvet-leaf blueberry, Labrador tea, alpine bilberry, and diapensia. Herbs include Rand's goldenrod, tussock bulrush, highland rush, mountain firmoss, mountain avens, mountain sandwort, and silverling. Some cliffs above 4,900 feet elevation lack temperate species, and some examples in the alpine zone contain only alpine-restricted species.

Montane - subalpine circumneutral cliffs occur in only a few locations in New Hampshire. They support calciphilic plant species absent from montane - subalpine acidic cliff communities, including northern cotton club rush, shrubby cinquefoil, livelong saxifrage, butterwort, and scirpus-like sedge. Many of these species require seasonally to perennially wet seepage zones with elevated levels of calcium. The montane - subalpine circumneutral cliff community is restricted to parts of large cliffs and ravines in the mountains.

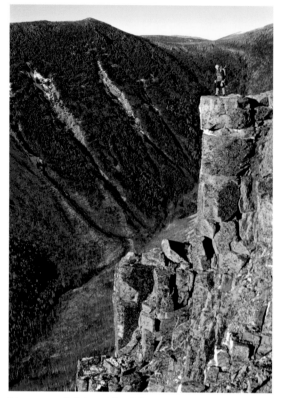

Montane - subalpine acidic cliff at Bondcliff in the White Mountains.

Montane - subalpine circumneutral cliff at Cannon Cliff in Franconia Notch.

Good Examples of Cliff Natural Communities

TEMPERATE CLIFFS

Temperate acidic cliff: Pawtuckaway State Park (Nottingham), Frankenstein Cliff and Harts Ledge (Hart's Location), and White Horse and Cathedral Ledges (Hale's Location/Bartlett).

Temperate circumneutral cliff: Cape Horn (Northumberland), Diamond Peaks (Second College Grant), Rattlesnake Mountain (Rumney), Holts Ledge (Lyme), and Whites Ledge (Bartlett).

MONTANE - SUBALPINE CLIFFS

Montane - subalpine acidic cliff: Cannon Cliff (Franconia), Whitewall Mountain (Bethlehem), Imp Face (Beans Purchase), Franconia Ridge (Franconia), and Tuckerman Ravine in the Presidential Range.

Montane - subalpine circumneutral cliff: Cannon Cliff (Franconia), Huntington Ravine (Sargents Purchase), and Butterwort Flume (Harts Location).

CHARACTERISTIC PLANTS OF
CLIFF NATURAL COMMUNITIES

A = Montane - subalpine acidic cliff   C = Temperate circumneutral cliff

B = Montane - subalpine circumneutral cliff   D = Temperate acidic cliff

COMMON NAME	SCIENTIFIC NAME	A	B	C	D
TREES					
Red spruce	Picea rubens	o	o	o	o
Heartleaf birch	Betula cordifolia	o	o		
Northern white cedar	Thuja occidentalis			o	
Paper birch	Betula papyrifera			o	o
Yellow birch	Betula alleghaniensis			o	o
Red oak	Quercus rubra			o	o
White pine	Pinus strobus			o	o
Eastern red cedar	Juniperus virginiana			o	o
SHRUBS					
Labrador tea	Ledum groenlandicum	o			
Mountain cranberry	Vaccinium vitis-idaea	o	o		
Diapensia*	Diapensia lapponica*	o	o		
Three-toothed cinquefoil	Sibbaldiopsis tridentata	o	o		
Meadowsweet	Spiraea alba	o	o	o	
Bush honeysuckle	Diervilla lonicera	o	o	o	o
Shrubby cinquefoil	Pentaphylloides floribunda		o	o	
HERBS					
Alpine bitter cress*	Cardamine bellidifolia*	o			
Mountain avens*	Geum peckii*	o	o		
Silverling*	Paronychia argyrocoma*	o	o		
Rand's goldenrod	Solidago simplex ssp. randii	o	o		
Highland rush	Juncus trifidus	o	o		
Large-leaved goldenrod	Solidago macrophylla	o	o		
Tussock bulrush	Trichophorum cespitosum	o	o		
Scirpus-like sedge*	Carex scirpoidea*		o		
Livelong saxifrage*	Saxifraga paniculata*		o		
Nodding saxifrage*	Saxifraga cernua*		o		
Common hairgrass	Deschampsia flexuosa	o	o	o	o
Early goldenrod	Solidago juncea	o	o	o	o
Fragile fern	Cystopteris fragilis	o	o	o	o
Rough bentgrass	Agrostis scabra	o	o	o	o
Harebell	Campanula rotundifolia		o	o	
Rusty woodsia	Woodsia ilvensis		o	o	
Maidenhair spleenwort	Asplenium trichomanes			o	
Slender cliffbrake*	Cryptogramma stelleri*			o	
Fragrant fern*	Dryopteris fragrans*			o	
Purple cliffbrake*	Pellaea atropurpurea*			o	
Early saxifrage	Saxifraga virginiensis			o	
Rock polypody	Polypodium virginianum			•	•
Colonial sedge	Carex communis			o	o
Marginal wood fern	Dryopteris marginalis			o	o
Silverrod	Solidago bicolor			o	o
BRYOPHYTES					
Calciphilic bryophytes			•	•	
Other bryophytes		•	•	•	•
Lichens		•	•	•	•

• = abundant to dominant   o = occasional or locally abundant   * = state threatened or endangered species

A **temperate lichen talus barren** above the Connecticut River on Wantastiquet
Mountain in Hinsdale.

Talus

Talus consists of jagged masses of rock that have accumulated on steep slopes below cliffs. The freeze-thaw action of water in cracks forces blocks of rock to separate and fall from the cliff. The most famous example of this phenomenon in New Hampshire occurred in 2003, when the formation known as the Old Man of the Mountain fell from its perch on Cannon Cliff onto the talus slope below. Talus slopes may be large or small, and are found throughout the state from sea level to the alpine zone. Talus is distinguished from rocky ridges and cliffs by the unique microenvironments within the sloping rock piles, and the high level of disturbance maintained by ongoing rockfall and shifting rock debris.

Talus slopes are extremely diverse physical environments, where rock size and slope stability, soil development, and climate vary within and among examples. Rock size and talus slope stability are a function of attributes of the parent cliff above, especially the degree of fracturing and frequency of freeze-thaw activity. At one extreme, cliffs with highly fractured rock types and frequent freeze-thaw action produce small, unstable talus fragments. Wet chasms or other weak, fractured sections of cliffs drop rocks regularly and produce zones on the talus below that are subject to frequent slides or avalanches. At the other extreme, cliffs with more massive, less fractured rock produce more stable talus with fewer, larger blocks. Rock size also varies from the top to the bottom of talus slopes. Smaller stones settle higher on the slope, and larger rocks tumble farther and collect at the base. Talus also occurs below bare cliff and outcrop areas of landslide tracks, typically on middle and lower portions of the slides.

Soil development on talus slopes is related to the frequency of disturbance, the size and spacing of rocks, and the amount of organic matter and sediment deposition. The top of the talus slope accumulates smaller stones, snow, runoff, fine sediments, leaves, and other organic matter that washes down from the cliff above, leading to greater soil development and plant growth. Talus comprised of large blocks is less hospitable for the establishment of plants because any leaves or other organic matter disappear into deep, dark cavities between the blocks. Lichens often dominate the rock surfaces in these areas. Similarly, unstable talus lacks vegetation or supports only a sparse cover of disturbance-tolerant species. More stable talus eventually accumulates soil and leads to the establishment of trees, shrubs, herbs, mosses, and liverworts. On some bare rock surfaces, lichens and mosses initiate the formation of soils and facilitate the growth of vascular plants.

On talus slopes, moisture, temperature, and light vary with latitude, aspect, and local physical features. South-facing slopes at lower latitudes and elevations are relatively warm and dry. Boulder tops are warmer and drier

Blackberries, meadowsweet, composites, vines, grasses, and tree saplings grow at the top of the Cannon Mountain talus slope, a collection zone for small talus fragments and organic debris sloughed from the cliff face above.

than their undersides, simply as a result of exposure. The upslope portion of a talus barren is warmer than the lower part due to sunlight reflecting from the cliff above. Northern, high-elevation areas are moister and colder than areas either farther south or at lower elevations. Colder, moister environments also occur on north-facing slopes, between boulders, and at the bases of talus slopes. The most dramatic cold-air microclimates occur among large boulders at the base of extensive talus slopes and in talus gorges. Late-melting ice beneath the boulders, and cooler air draining downslope, create a moist, cold environment. These cold microclimates support montane conifers and subalpine species in close proximity to vegetation indicative of a much warmer climate. Good examples of this phenomenon can be observed at Ice Gulch in Randolph and Zealand Notch in Bethlehem, where subalpine plants occur at low elevations, and at Mount Wantastiquet in Chesterfield, where red spruce and mountain ash occur much farther south and at lower elevation than usual. Many of these cold-air talus environments are stable, "fossil talus" slopes that no longer actively accumulate new rock. Fossil talus consists of giant boulders that tumbled to their present locations during post-glacial, cold-climate periods when freeze-thaw activity was greatest.

Trees, shrubs, vines, herbs, bryophytes, and lichens may all be present on talus, but their distribution varies greatly with microsite conditions. Trees and shrubs grow where rockfall and slides are infrequent enough to allow their establishment and growth on soil located between or on top of boulders. Wooded talus slopes range from sparse woodland to woodland. Conifers and heath shrubs occur in cold-air talus environments, on north-facing talus slopes and deep gorges, and above 3,000 feet elevation. Deciduous trees, especially birches, maples, and oaks, dominate talus slopes in warmer climate settings. Shrubs and vines are frequent here as well. Mountain maple is almost always present on talus slopes, forming thickets in some areas. Gooseberries and currants are also common. Vines such as Virginia creeper and fringed bindweed trail over and among boulders. Mosses, liverworts, and ferns such as rock polypody and marginal woodfern adorn large, stable boulders in cold, moist environments. Herbs grow where finer sediments and organic matter accumulate between boulders, though disturbance-tolerant herbs occur in some unstable talus areas. Lichens are the dominant life-form on large or unstable block talus where little soil has accumulated.

A variety of adaptations allow plants of talus slopes to survive the stressful conditions. In nutrient-poor, patchy soils, plants emphasize root growth to enhance uptake of soil resources. Stems that can reorient following disturbance—along with deep, spreading, pliable roots—allow certain grasses and forbs to tolerate actively shifting talus environments. Annual plants complete their life cycle in one growing season and overwinter as seeds, allowing them to minimize their exposure to rockfall. Vines root in patchy

Soils accumulate slowly on talus slopes and support plants, such as rock polypody, that are adapted to the acidic, nutrient-poor, and seasonally dry conditions.

Herb Robert, a common herb of talus slopes, at Oak Hill in Tuftonboro.

soil, grow over boulders, and spread into canopy gaps. Plants that grow directly on dry, open rock, such as rock polypody and marginal wood fern, tolerate desiccation. Mosses and liverworts are also well adapted to boulder tops. Lichens are especially well adapted to grow on open rock and are sometimes the only form of life present.

Talus Barrens, Shrublands, and Landslides

Talus barrens are mostly treeless, open slopes with large, lichen-covered rocks and little or no soil accumulation or vascular plant cover. Crustose, umbilicate, and foliose lichens are the dominant life-forms. Plants abundant

Montane lichen talus barren on Bondcliff.

Temperate lichen talus barren at Ellis Hatch Wildlife Management Area in Brookfield.

in adjacent wooded talus or forest communities, including mountain maple, raspberries, currants, grasses, and rock polypody, are sparse in talus barren communities. Talus barrens in landslide tracks require decades to establish lichens, and centuries to accumulate enough soil to support sparse woodlands. **Montane lichen talus barrens** occur mostly above 2,200 feet elevation, and occasionally as low as 1,500 feet. Scattered spruce and birch trees may be present, but are often stunted from exposure. **Temperate lichen talus barrens** occur below 1,800 feet elevation in the White Mountains and central and southern New Hampshire. Scattered oak and birch trees may be present.

Subalpine cold-air talus shrubland is a rare community found at the base of large talus slopes in the White Mountains. Cool air emerges from

Subalpine cold-air talus shrubland at Ice Gulch in Randolph.

Montane landslide barren and thicket in the Pemigewasset Wilderness.

Spruce - moss wooded talus at Devil's Hopyard in Stark.

beneath boulders where ice lingers into late summer. The cold microenvironment has a profound effect on vegetation, creating conditions suitable for subalpine plants otherwise restricted to higher-elevation mountaintops. Dwarf red and black spruce reach less than 5 feet in height, and a short layer of shrubs grows with montane lichens on the surface of talus boulders. Characteristic species include alpine bilberry, purple crowberry, Labrador tea, and sheep laurel.

Montane landslide barren and thicket communities are found on debris deposit areas and exposed till of the lower portions of landslide tracks (below outcrops, cliffs, and talus). Vegetation succeeds through barren and thicket stages and eventually becomes forest. Pioneer species include sedges, grasses, rushes, hay-scented fern, whorled aster, other weedy forbs, willows, alder, and mosses. Birch, aspen, and pin cherry seedlings may also be present. Forest redevelopment is slow on steep, eroded till areas. Birch, spruce, and fir saplings are typically stunted at no more than a foot high for at least 20 years. The most rapid successions occur in deposit areas where pioneer hardwoods completely crowd out light-dependent pioneer herbs within 30 years. Shade-tolerant forest herbs may take up to 70 years to recover after a landslide.

Birch - mountain maple wooded talus at Mount Stanton in Bartlett.

Wooded Talus

Spruce - moss wooded talus forms in cold environments, including north-facing slopes and gorges, and immediately downslope from subalpine cold-air talus shrublands. Extensive moss and liverwort mats cover large boulders beneath an overstory of black and red spruce, balsam fir, and an occasional birch. Creeping snowberry, Labrador tea, velvet-leaf blueberry, northern wood sorrel, bluebead lily, and rock polypody are the dominant shrubs and herbs. Ice remains beneath the boulders until late spring or early summer. Most examples of spruce - moss wooded talus occur above 2,300 feet elevation, but disjunct occurrences can be found as low as 800 feet.

Birch - mountain maple wooded talus communities range from sparse woodland to nearly forested in structure. Trees are less than 30 feet in height, and shrubs and herbs are scattered or locally abundant among the lichen-dominated rocky openings. This community primarily occurs in central and northern parts of the state above 1,600 feet, and extends above 3,000 feet in places. Northern birches (yellow, paper, and heartleaf), pioneer maples (mountain, striped, and red), and mountain ash are the primary trees and shrubs. Red spruce is common, but is relatively low in abundance. Bluebead lily, northern wood sorrel, and other ferns and herbs of northern forests are present. Moss cover is less dense than in spruce - moss wooded talus.

Red oak - black birch wooded talus communities occur below 1,600 feet in central and southern New Hampshire. Red oak is the dominant tree species, and is more abundant than the northern birches, maples, and spruce characteristic of birch - mountain maple wooded talus. Black birch is present, a species notably absent in higher-elevation talus communities. Dry-site species, including ironwood, goldenrods, sedges, grasses, blueberries,

CHARACTERISTIC PLANTS OF TALUS NATURAL COMMUNITIES

A = Montane *and* temperate lichen talus barrens
B = Subalpine cold-air talus shrubland
C = Spruce - moss wooded talus
D = Birch - mountain maple wooded talus
E = Red oak - black birch wooded talus

COMMON NAME	SCIENTIFIC NAME	A	B	C	D	E
TREES						
Red spruce	*Picea rubens*	o	•	•	o	
Black spruce	*Picea mariana*		•	o		
Balsam fir	*Abies balsamea*	o	•	•		
Heartleaf birch	*Betula cordifolia*	o	•	o	•	
Paper birch	*Betula papyrifera*			o	•	o
Yellow birch	*Betula alleghaniensis*	o			•	•
Red maple	*Acer rubrum*				•	•
Red oak	*Quercus rubra*	o				•
Gray birch	*Betula populifolia*	o				o
Sugar maple	*Acer saccharum*					•
American beech	*Fagus grandifolia*					•
Black birch	*Betula lenta*	o				•
Ironwood	*Ostrya virginiana*					•
SHRUBS & VINES						
Alpine bilberry	*Vaccinium uliginosum*		•			
American mountain ash	*Sorbus americana*		•	o	•	
Mountain cranberry	*Vaccinium vitis-idaea*		•			
Labrador tea	*Ledum groenlandicum*		•	•		
Creeping snowberry	*Gaultheria hispidula*	o	•			
Velvet-leaf blueberry	*Vaccinium myrtilloides*		•	•	•	
Lowbush blueberry	*Vaccinium angustifolium*		•	•	•	•
Mountain maple	*Acer spicatum*	o		o	•	•
Gooseberries and currants	*Ribes* spp.	o			•	•
Striped maple	*Acer pensylvanicum*				•	•
Virginia creeper	*Parthenocissus quinquefolia*				•	•
Fringed bindweed	*Fallopia cilinodis*				•	•
Raspberries and blackberries	*Rubus* spp.				•	•
Witch hazel	*Hamamelis virginiana*					•
Maple-leaved viburnum	*Viburnum acerifolium*					•
HERBS						
Northern wood sorrel	*Oxalis montana*			•	•	
Bluebead lily	*Clintonia borealis*			•	•	
Rock polypody	*Polypodium virginianum*	o		•	•	•
Common hairgrass	*Deschampsia flexuosa*	o			•	o
Pale corydalis	*Corydalis sempervirens*	o			o	o
Marginal wood fern	*Dryopteris marginalis*				•	•
False Solomon's seal	*Maianthemum racemosum*					•
Distant sedge	*Carex lucorum*					o
Rough-leaved rice grass	*Oryzopsis asperifolia*					o
Bracken	*Pteridium aquilinum*					o
Goldenrods	*Solidago* spp.					o
Wild sarsaparilla	*Aralia nudicaulis*					o
NON-VASCULAR						
Lichens		•	•	•	•	•
Bryophytes		o	•	•	•	o

• = abundant to dominant   o = occasional or locally abundant   * = state threatened or endangered species

maple-leaved viburnum, and witch hazel, are more prominent here than in more northern and higher-elevation talus communities.

Appalachian wooded talus is a rare community found on low-elevation, acidic slopes in southern and coastal parts of the state. The characteristic species of this community—Appalachian oaks (white, black, etc.), mountain laurel, early azalea, dogwoods, and sassafras—do not occur in other talus communities.

Good Examples of Talus Natural Communities

TALUS BARRENS, SHRUBLANDS, AND LANDSLIDES

Montane lichen talus barren: Below Cannon Cliff (Franconia), Ice Gulch (Randolph), Carter Notch (Beans Purchase), King and Castle ravines in the Presidential Range, and Zealand Notch (Lincoln).

Temperate lichen talus barren: Mount Monadnock (Jaffrey), Ossipee Mountains (Ossipee), Pawtuckaway State Park (Nottingham), Ellis Hatch Wildlife Management Area (Brookfield), and Sundown Ledge (Albany).

Montane landslide barren and thicket: Common on many steep slopes in the White Mountains.

WOODED TALUS

Subalpine cold-air talus shrubland: Below Cannon Cliff (Franconia), Ice Gulch (Randolph), Carter Notch (Beans Purchase), and King and Castle ravines in the Presidential Range.

Spruce - moss wooded talus: Below Cannon Cliff (Franconia), Ice Gulch (Randolph), Carter Notch (Beans Purchase), King and Castle ravines in the Presidential Range, Magalloway Mountain (Pittsburg), and Devil's Hopyard (Stark).

Birch - mountain maple wooded talus: Cape Horn (Northumberland), Ice Gulch (Randolph), King and Castle ravines in the Presidential Range, Mount Hale (Bethlehem), Mount Monadnock (Jaffrey), and the Ossipee Mountains (Ossipee).

Red oak - black birch wooded talus: Pawtuckaway State Park (Nottingham), Ellis Hatch Wildlife Management Area (Brookfield), Sundown Ledge (Albany), and Mount Stanton (Bartlett).

Appalachian wooded talus: Mount Wantastiquet (Hinsdale).

Rocky Ground Wildlife

The unique physical and geologic characteristics of rocky ground communities support an unusual diversity of wildlife habitats. Many species use rock cavities for denning or for protection from predators, seek out sunny openings for basking and thermal regulation, and forage or hunt for food among the sparse vegetation. Ridges and south-facing cliffs and talus receive ample sun and offer habitat for wildlife species that thermoregulate or are intolerant of cold, such as snakes and invertebrates, and plentiful cracks and crevices for other animals to hide or nest in.

Rocky ground provides many common wildlife species with short term, but essential, habitats. For example, crevices in rock outcrops attract porcupine, black bear, gray fox, and fisher for denning or temporary refuge. Bobcats use warm, southwest-facing rocky slopes that offer secure loafing sites and relief from cold temperatures and deep snow. Wooded rocky ridges function as important movement corridors for large, wide-ranging mammals like moose, black bear, and white-tailed deer.

Several snake species use outcrops associated with rocky ridges for basking and denning, including the eastern garter snake, black rat snake, northern redbelly snake, and the nearly extirpated timber rattlesnake. Males and females of some species bask until their skin sheds, then move into forested habitats for the summer. Pregnant females remain near exposed slopes and protective rocks until they are ready to bear young, usually in September.

A peregrine falcon chick and eggs on a ledge on a **montane - subalpine acidic cliff** at Abeniki Mountain in Dixville.

Some snakes winter in communal dens called hibernacula, located within underground crevices below the frost line. The snakes emerge from the hibernacula in spring and seek exposed rocky areas where they alternately bask and seek shelter from the sun.

Peregrine falcons are a conspicuous wildlife species of cliffs. They nest in shallow scrapes or small caves on exposed, near-vertical cliff faces. From these steep ledges, peregrines hunt for food, striking and killing birds in mid-air before swooping to the ground to retrieve the prey. Cliff nests offer ideal protection from predators. Peregrine falcons are a threatened species in New Hampshire, driven nearly to extinction in the 1980s from DDT poisoning, which resulted in eggs so brittle that they broke under the weight of brooding adults. Populations have rebounded in recent years, however, reaching a high of seventeen active pairs in the state in 2008.

Several bat species use rocky crevices in cliff communities as daytime summer roosting sites that provide thermally stable environments as well as protection from the elements and predators. The rare eastern small-footed bat can tolerate colder temperatures than other species and forms maternal colonies in rocky crevices on steep, mountain cliff faces. Female bats spend summer days resting and tending to their young. Other bats that use crevices in cliffs for roosting include the eastern pipistrelle and little brown bat. Bats in the Northeast are currently under threat by white nose syndrome, a fungus spreading rapidly among hibernacula.

The crevices and jumbled rock of talus slopes also provide habitat for wildlife. In wooded talus slopes, the combination of well-drained soils, deep leaf litter, and varied groundcover is ideal for small mammals such as rock voles and long-tailed shrews. These species use the rocky terrain for foraging, denning, and protection from predators. Snakes take advantage of the abundance of small mammal prey, ample denning sites, and the presence of open sunny areas for basking. New Hampshire's talus slope communities also harbor many spiders, insects, and land snails.

3 Forests

New Hampshire's forests encompass features as varied as spires of spruce and fir in the North Country, towering white pines in southern parts of the state, and the brilliant fall foliage of hardwoods throughout. At present, forests cover more than 80 percent of the state, in contrast to the dominance of open pasture and farmland as recently as 150 years ago. Trees rapidly reclaimed the landscape after settlers abandoned farmland to move to urban areas or migrate west. Today, the state's diverse, resilient, and plentiful forests are important to its environmental and economic health, and to the quality of life of its residents. They provide critical ecosystem services such as carbon storage, flood storage, groundwater recharge, and pollutant removal, and support wood product, recreation, and tourism industries.

Forests are the most common terrestrial ecosystem worldwide, occurring wherever temperature, moisture, and disturbance are suitable. Varying climatic conditions in continental North America yield several forest regions, each distinguishable by vegetation, soils, and ecosystem processes. Three of these forest regions occur in New Hampshire.

Acadian spruce - fir forests dominate cold-climate conifer forests of the northern Appalachian Mountains, adjacent portions of northern New England and maritime Canada, and the Adirondacks. They are similar to the northern boreal forests that encircle the polar region, but experience cooler and wetter summers, have greater tree diversity, and less frequent fires. In New Hampshire, spruce - fir forests occur at high elevations and in cold, lowland valleys in the mountains and North Country. Red spruce and balsam fir dominate the forest canopy, birches are important successional

PRECEDING PAGE: The Big Pines Natural Area at Hemenway State Forest in Tamworth.

Stone walls like this one in Hollis reflect the agricultural legacy of many of New Hampshire's forests.

species, and dwarf shrubs and bryophytes are abundant in the understory. Plants of Acadian spruce - fir forests tolerate seasonal temperature extremes, and compete for scarce nutrients during the short growing season.

Laurentian mixed forests are cool-temperate forests occupying a broad east-to-west zone along the U.S. and Canadian border, from the northern Great Lakes states to the Maritime provinces, including the mountainous terrain of northern New England. They cover most of New Hampshire's land area between the Acadian spruce - fir forests of the north and mountains and the Appalachian oak and pine forests of southern and coastal parts of the state. Laurentian mixed forests contain hardwood and conifer trees, including many species absent or sparse in other New Hampshire forests. Sugar maple, American beech, and yellow birch are the dominant species in more northern and higher-elevation areas. Hemlock, beech, red oak, and white pine are abundant in southern and lower-elevation areas. Tall shrubs and herbs are common in the understory. Many Laurentian mixed forest plants require warmer and longer growing seasons, and more productive soils, than plants of Acadian spruce - fir forests.

Appalachian oak and pine forests are warm-temperate forests of the central and Appalachian states. They extend into the southern Great Lakes states and central New England, and reach their northeastern limit in southern and coastal New Hampshire. The dominant hardwood trees are red oak, white oak, black oak, and hickories. Pitch and white pine are the dominant conifers. Hemlock, sugar maple, and yellow birch are sparse. Shrubs and herbs are common in the understory. Many Appalachian oak and pine forest plants are drought resistant and fire tolerant, adaptations to the warm, dry climate and site conditions. Fire was more frequent in Appalachian oak and pine forests prior to the enactment and enforcement of fire-suppression policies.

Rich woods are a special subset of hardwood forest natural communities within the Laurentian and Appalachian forests. These communities share a diverse assemblage of plants restricted to nutrient-rich conditions. Sugar maple and white ash are the most common canopy trees, while maidenhair fern, silvery spleenwort, baneberries, Dutchman's breeches, and blue cohosh characterize a lush understory. Many of New Hampshire's rare forest plant species occur in rich woods communities.

The development of New Hampshire's current forests began about 10,000 years ago, following a period when tundra vegetation dominated the landscape after glacial retreat. Individual tree species arrived at different times, reflecting differences in dispersal modes and rates, migration routes, geographic barriers, and distances from glacial refugia. Wind-dispersed conifers and aspens arrived first, followed by American elm and white pine 1,000 years later. Hemlock arrived 8,000 years ago, American beech 6,500 years

Forest damage from a tornado in central New Hampshire.

ago, hickory 5,000 years ago, and American chestnut 2,000 years ago. By that time, species ranges had stabilized, and resembled vegetation patterns of today. Forest composition will likely continue to respond to changes in climate over the coming centuries and millennia.

Individual forest communities vary greatly in size. A few common communities occupy extensive areas, forming regional forest matrices. A greater number of communities occur as smaller patches embedded within the forest matrix, corresponding to sites with different microclimates, site conditions, and disturbance regimes. The communities that form the forest matrix of one region, however, may only occur as patches in adjacent regions. For example, spruce - fir forests, common in the White Mountains, occur in southern New Hampshire only as isolated patches on summits like Mount Monadnock.

Disturbances affecting New Hampshire forests vary in frequency, intensity, and scale. Occasional hurricanes and other major storms cause canopy damage and blowdowns over large areas. Fungi, insects, wind, and small storms regularly impact canopies and create small gaps in the forest. Fungi and other pathogens weaken trees, and increase their susceptibility to wind and storm damage. Insect pests can harm trees through defoliation or attacking vascular tissues. Herbivorous mammals such as deer, moose, and porcupines browse young trees and strip bark from stems and branches, increasing susceptibility to insects, fungi, and other pathogens. Fire is infrequent in New Hampshire forests, but occurs locally on dry soils in central and southern parts of the state. Fire was historically important for the maintenance of certain forest types.

Disturbances that form gaps in the forest canopy alter amounts of light, nutrients, and moisture, which in turn affect the competitive balance among species. Light is the primary resource in forests; young plants must either tolerate shade or colonize openings in the canopy when trees die or fall. Gap colonizers such as paper birch, white ash, and red maple are fast-growing, nutrient-demanding, short-lived species that produce wind-dispersed seeds. In time, these gap colonizers give way to slower-growing, shade-tolerant, longer-lived trees. Most forest herbs are shade-tolerant, but some are shade-intolerant and rely on gap formation for survival. Plants already established in the understory have a distinct advantage over new colonizers when a canopy gap forms.

Humans directly and indirectly affect the composition of New Hampshire's forests. The abundance of late successional, shade-tolerant trees such as red spruce, hemlock, and beech has decreased in the state due to timber harvesting and historic clearing for agriculture. Shade-intolerant species such as white pine, paper birch, aspen, and cherry have increased, in some instances obscuring relationships between late successional trees and climate or site conditions. Non-native, invasive species have displaced native plants in some forest canopies and understories. Oriental bittersweet, glossy buckthorn, and Japanese barberry are a few of the invasive species that can out-compete native plants. This phenomenon is particularly common in forest patches of fragmented landscapes and along roadsides. In addition, deer populations also thrive in fragmented habitats, particularly in suburban settings where hunting pressure is low. In some areas, browsing affects the herbaceous layer and tree regeneration severely enough to change the character of the forest.

Exotic insects, pathogens, pollution, and climate change pose significant and immediate threats to forests in the region. For example, the hemlock woolly adelgid, an insect from Japan, damages and kills hemlock trees. Accidental introduction of the fungus *Diaporthe parasitica* to North America in the early 1900s eliminated mature American chestnut by 1940. Fossil fuel combustion produces acid precipitation, which chronically affects forest nutrient cycling and species composition. Spruce - fir forests are especially susceptible to acid rain. Global warming is altering forest ecosystem processes and composition, inducing a northward and upslope migration of forest species and natural communities.

FOREST NATURAL COMMUNITIES (NORTHERN NEW HAMPSHIRE)

PARENT MATERIAL

Colluvium on till or talus	Alluvium (outwash, kame, lakebed sediments)	Compact basal till	Fine, firm to loose ablation till	Shallow or very rocky till

NATURAL COMMUNITIES

A or B	C or D	D or E	F	E	G	H

A. **Rich mesic forest** – Colluvial slopes and benches, and areas influenced by calcium-rich bedrock (also occurs in southern New Hampshire)

B. **Semi-rich mesic sugar maple forest** – Colluvial slopes or fine till soils (also southern NH)

C. **Lowland spruce - fir forest** – Valley bottoms on alluvial or gently sloped till soils below 2,500 ft.; most common in the North Country

D. **Hemlock - spruce - northern hardwood forest** – Valley bottoms and stream drainages on alluvial or compact till soils below 2,000 ft.; most common in the White Mountains

E. **Northern hardwood - spruce - fir forest** – Mostly above 2,500 ft.; occasional in the White Mountains and common in the North Country at lower elevations on compact till or rocky soils

F. **Sugar maple - beech - yellow birch forest** – Fine ablation till below 2,500 ft.

G. **High-elevation spruce - fir forest** – Mostly between 2,500 and 4,000 ft., locally higher in protected sites, and lower on wind-exposed ridges or rocky sites

H. **High-elevation balsam fir forest** – Mostly between 4,000 and 4,900 ft., locally lower on wind-exposed ridges

FOREST NATURAL COMMUNITIES
(CENTRAL AND SOUTHERN NH)

PARENT MATERIAL

Coarse rocky till	Fine, mesic till (basal or ablation) or colluvium	Shallow or rocky, dry till	Mesic or dry-mesic till (basal or ablation)	Mesic or dry-mesic alluvium	Wet alluvium	Dry, coarse alluvium (sand and gravel)

NATURAL COMMUNITIES

Central New Hampshire below 2,000 ft. elevation						
A or B	C or D	E	B or F		G	H or I

Southern New Hampshire below 1,000 ft. elevation						
A	J or K	L	F or K		G	H

A. **Hemlock forest** – Mostly in rocky ravines and on shallow rocky slopes

B. **Hemlock - oak - northern hardwood forest** – Rocky slopes and till soils up to 2,000 ft.

C. **Semi-rich mesic sugar maple forest** – Fine, mesic till soils and colluvial sites

D. **Sugar maple - beech - yellow birch forest** – Fine, mesic till soils mostly above 1,500 ft.

E. **Dry red oak - white pine forest** – Shallow, rocky till soils between 1,000 and 2,000 ft.; also on sand plains

F. **Hemlock - beech - oak - pine forest** – Very common in central and southern New Hampshire on mesic and dry-mesic till and alluvial soils; mostly below 1,500 ft.

G. **Hemlock - cinnamon fern forest** – Till or alluvial soils with a seasonally high water table

H. **Pitch pine - scrub oak woodland** – Fire-maintained community on coarse sand plain soils

I. **Mixed pine - red oak woodland** – Fire-maintained community on coarse sand plain soils in central New Hampshire

J. **Semi-rich oak - sugar maple forest** – Dry-mesic colluvial till or talus soils

K. **Mesic Appalachian oak - hickory forest** – Mesic to dry-mesic soils below 800 ft., mostly in coastal New Hampshire

L. **Dry Appalachian oak forest** – Shallow, rocky till and sand plain soil below 1,000 ft.

Old-growth forest at the Nancy Brook Research Natural Area.

Old-Growth Forests

The forest that greeted New Hampshire's first European settlers exists only as scattered remnants today. These remaining patches, known as old-growth, virgin, primeval, or ancient forests, have escaped harvesting or other significant human modification over the last 350 years, and cover less than a tenth of one percent of the state.

Strictly defined, the term "old growth" means a forest that has never been cut or altered, though some second-growth or regenerating forests can also develop old-growth characteristics if sufficient time has passed to obscure the effects of human disturbance. In the northeast, at least 200 years is required to develop an old forest structure, although some old-growth traits begin to develop after 100 years. Due to the difficulty in identifying true old-growth with certainty, we use the term "old forest" more loosely to describe stands that have many of the characteristics one would expect in an old-growth forest, such as old age, complex canopy structure, and coarse woody debris. Whether truly old-growth or not, old forests are important and unique ecological features that warrant our attention and protection.

Trees in New Hampshire's old forests typically have long trunks free of lower branches, deeply furrowed or plated bark, heartwood decay, large and prominent roots, and thick limbs. Snags and downed logs in all stages of decomposition are abundant. Typical canopy species include red spruce, hemlock, yellow birch, beech, sugar maple, and black gum. Most of the tree species are shade-tolerant, although blowdowns also create opportunities for gap colonizers such as paper birch and quaking aspen. Beneath the canopy, the forest floor undulates with pits and mounds where trees have fallen and decomposed.

New Hampshire's old-growth forests are valuable and endangered natural ecosystems. They provide habitat for a wide variety of animals, microorganisms, and flowering plants. Additionally, many mosses, lichens, and fungi depend on these forests. Scientists derive substantial knowledge about forest structure and function from studies of old growth. Although tree girth in New Hampshire's old forests may be smaller than one might expect, one can't help noticing that something remarkable exists in the nature of these old stands.

Acadian spruce - fir forest in the White Mountain National Forest.

Acadian Spruce - Fir Forests

Acadian spruce - fir forests are conifer-dominated woods of cold climate areas in northern New England, New York, and the Canadian Maritime provinces. In New Hampshire, they occur at high elevations and high latitudes, cloaking the upper slopes of mountains and filling many low-lying valleys and basins in the North Country and White Mountains. Cold temperatures, moisture, nutrient-poor soils, and a short growing season in these settings lend an advantage to conifers over hardwoods.

The vegetation of most Acadian spruce - fir forests is remarkably consistent. Red spruce and balsam fir are abundant at both high and low elevations. Yellow, paper, and heartleaf birch are the primary colonizers of gaps created by fallen trees. Striped maple and mountain ash are common in the understory. Black spruce and low heath shrubs occur on wet lowland sites where drainage is poor, and near treeline where the climate is severe. White spruce occurs north of the White Mountains, and white pine is occasional in lowland examples of these forests. Mosses, lichens, and dwarf shrubs such as bunchberry and creeping snowberry are common in the moist understory, along with a few herbaceous plant species such as northern wood sorrel and mountain wood fern. These communities are less diverse than hardwood forests to the south, but more diverse than boreal forests to the north. In total, about 280 plant species occur in Acadian spruce - fir forests in New Hampshire.

Acadian spruce - fir forests are frost-free for 90 to 120 days per year, about 30 days less than northern hardwood forests. The colder climate that induces a shift from deciduous to coniferous forests is associated with high eleva-

Conifer tree spire in winter on the upper slope of Mount Moosilauke.

Spruce - fir forest in northern New Hampshire.

Bluebead lily and bunchberry are common plants in spruce - fir forests.

Interior of a fir wave on Mount Adams.

tions and lowland valley settings at northern latitudes. In New Hampshire, air temperature decreases 5 degrees for every 1000 feet of elevation gain, equivalent to a shift north of 200 miles. In addition, valley bottoms are disproportionately colder and wetter than mountain slopes. Cold air is dense and settles in the lowlands, where soils are also often wet, poorly drained, and take longer to warm in the spring. Each of these factors lends an advantage to conifers over hardwoods. This results in the characteristic pattern of spruce - fir forests growing on both mountaintops and valley bottoms in northern New Hampshire, with deciduous hardwood forests in intermediate settings.

Conifers of Acadian spruce - fir forests are well adapted to the cold climate. Most conifer trees are evergreens with firm, waxy needles retained for more than one year. This conserves nutrients and promotes rapid photosynthesis in the spring. Conifers commit less nitrogen and other nutrients to their needles than hardwoods do to their leaves, allowing them to invest proportionally more resources in their roots and maximize nutrient uptake from the poor soils. Leaves low in nutrients are high in decay-resistant, carbon-rich compounds like lignin. When shed, the needles decompose slowly, contributing to the accumulation of soil organic matter. In addition, the firm needles and narrow, cone-shaped crowns of conifer trees shed heavy snow and ice, allowing them to minimize damage during long, harsh winters. Tall, narrow crowns also help conifers expose more leaf area to the sun, which strikes the earth at a low angle in northern latitudes.

Elevation and landscape setting affect spruce - fir forest community disturbance patterns and species composition. Wind is an important disturbance agent in all spruce - fir forests, and is most important at high elevations on ridges with northern and western aspects. Tree canopies here are shorter, more sculpted, and lower in species richness than canopies at lower elevations. Chronic wind stress produces short-interval cycles of windthrow and forest succession, favoring short-lived and cold-tolerant balsam fir and heartleaf birch. Blowdowns in high-elevation fir forests sometimes occur as arc-shaped zones of dead snags and progressively older regeneration called fir waves. From a distance, they form an undulating, wave-like pattern of gray snags and green fir trees across upper slopes of mountainsides.

Spruce - fir forests at lower elevations suffer less wind stress, experience longer disturbance-free intervals, and have taller, more diverse tree canopies than higher-elevation forests. Insects and fungal pathogens are important disturbance agents at lower elevations. Red spruce is more common at lower elevations, and occurs with paper and yellow birch, while heartleaf birch is less common at low elevations than at high elevations. In addition, lowland spruce - fir forests have white spruce, and occasionally white pine, both of which are absent at higher elevations.

High-elevation balsam fir forest on Mount Bond.

Shade-intolerant birches are the primary gap colonizers in spruce - fir forests. Light, wind-dispersed seeds aid in dispersal to the gaps, where they regenerate well on bare mineral soil exposed by blowdowns. Paper and heartleaf birches grow fast and quickly fill gaps, but rarely live more than 130 years. In contrast, yellow birch and red spruce are relatively shade-tolerant gap colonizers. These species can live up to 400 years, surviving in the understory until a gap arises, and then persisting as canopy trees. Balsam fir is a more broadly adapted species of spruce - fir forests. It regenerates on organic duff or mineral soil, grows rapidly, is more shade-tolerant than birches, and can persist in the understory until a gap is formed. These attributes make it both a rapid gap colonizer and a shade-tolerant opportunist.

High-Elevation Mountain Slope Forests

High-elevation balsam fir forests occur on upper mountain slopes generally above 4,000 feet. It can also occur as low as 3,500 feet on wind-exposed ridges. Balsam fir dominates, progressively diminishing in height as elevation increases. At treeline, it transitions to krummholz and alpine communities. Other trees and tall shrubs include heartleaf birch, black spruce, and mountain ash. Red spruce, yellow birch, and paper birch are notably absent. Herbs sometimes form dense, glade-like understories in areas with deep snowpacks or broken tree canopies. Wind stress is more severe in this forest community than in any other type, resulting in many blowdown gaps and fir waves.

High-elevation spruce - fir forests occur below high-elevation balsam fir forests on mountain slopes from 2,500 to 4,000 feet elevation. They also

High-elevation spruce - fir forest in Great Gulf.

107

Lowland spruce - fir forest at Pondicherry Wildlife Refuge in Jefferson.

can occur below 2,500 feet, covering smaller areas on shallow, rocky soils of ridgelines and talus slopes, and on north-facing slopes in the North Country. Red spruce and balsam fir are the dominant trees. Yellow, paper, and heartleaf birch are common gap colonizers, along with shrub and short tree species such as striped and mountain maple, and mountain ash. The understory consists of bunchberry, northern wood sorrel, twinflower, bluebead lily, creeping snowberry, and wood ferns. A lush growth of mosses and liverworts blankets the ground, and lichens often cover trees and boulders.

Lowland Valley Bottom Forests

Lowland spruce - fir forests occupy valley bottoms below northern hardwood forests in northern New Hampshire. They form on moderately well to somewhat poorly drained soils in former lake basins, along stream drainages, and adjacent to swamps and peatlands. Soils derive from various parent materials including compact and loose glacial tills, and water-deposited lake-bottom, river, and kame terrace sediments. Red spruce, balsam fir, mosses, and lichens are abundant, and a variety of herbs and shrubs may be present. Lowland spruce - fir forest is floristically similar to high-elevation spruce - fir forest, but differs in landscape setting, soil drainage, and wind-stress characteristics. Some tree species present in lowland forests, but absent in their high-elevation counterpart, include white spruce, black spruce, and white pine.

Montane black spruce - red spruce forest is an uncommon to rare community type in New Hampshire. It occurs on somewhat poorly drained soils around heath woodlands and fens in high-elevation valleys above 2,000 feet in the White Mountains and North Country. Black spruce and red spruce

CHARACTERISTIC PLANTS OF SELECTED ACADIAN SPRUCE - FIR FOREST NATURAL COMMUNITIES

A = High-elevation spruce - fir forest   C = Montane black spruce - red spruce forest
B = High-elevation balsam fir forest   D = Lowland spruce - fir forest

COMMON NAME	SCIENTIFIC NAME	A	B	C	D
TREES					
Red spruce	*Picea rubens*	•	o	•	•
Black spruce	*Picea mariana*	•	o	•	
Balsam fir	*Abies balsamea*	•	•	o	•
Paper birch	*Betula papyrifera*	•	o		o
Yellow birch	*Betula alleghaniensis*	o			o
Heartleaf birch	*Betula cordifolia*	o	o		o
White spruce	*Picea glauca*				o
Quaking aspen	*Populus tremuloides*				o
Larch	*Larix laricina*				o
SHRUBS					
Bartram's serviceberry	*Amelanchier bartramiana*	o		o	o
Mountain holly	*Nemopanthus mucronatus*	o	o		o
Velvet-leaf blueberry	*Vaccinium myrtilloides*	o	o		o
Showy mountain ash	*Sorbus decora*	o	o		o
Sheep laurel	*Kalmia angustifolia*			o	
Labrador tea	*Ledum groenlandicum*			o	
Bunchberry	*Cornus canadensis*	o		o	o
Creeping snowberry	*Gaultheria hispidula*	o	o	o	o
Striped maple	*Acer pensylvanicum*	o			o
HERBS					
Intermediate wood fern	*Dryopteris intermedia*	o			o
Mountain wood fern	*Dryopteris campyloptera*	o	o		o
Northern wood sorrel	*Oxalis montana*	o	o	o	o
Bluebead lily	*Clintonia borealis*	o	o	o	o
Starflower	*Trientalis borealis*	o	o		
Twinflower	*Linnaea borealis*		o		o
Goldthread	*Coptis trifolia*	o	o		o
Canada mayflower	*Maianthemum canadense*	o	o		
Shining clubmoss	*Huperzia lucidula*	o			
Cinnamon fern	*Osmunda cinnamomea*			o	
Long beech fern	*Phegopteris connectilis*	o			
Wild sarsaparilla	*Aralia nudicaulis*				o
Wakerobin	*Trillium erectum*				o
Foamflower	*Tiarella cordifolia*				o
Heart-leaved twayblade*	*Listera cordata**	o	o		o
Northern comandra*	*Geocaulon lividum**	o	o		

• = abundant to dominant   o = occasional or locally abundant   * = state threatened or endangered species

Northern white cedar forest/woodland is an uncommon community of northern New Hampshire.

mix in various combinations as the dominant trees. Shrub, herb, and bryophyte species compositions resemble those of other spruce - fir forests.

Northern white cedar forest/woodland is an uncommon upland forest community limited to low-elevation settings in the North Country. Dominated by northern white cedar, the community occurs on rocky knobs and river terrace slopes. Spruce, fir, and hardwoods are present in lesser abundance. Understory vegetation is sparse, and includes wild sarsaparilla, intermediate wood fern, and starflower. Northern white cedar forest/woodland is noteworthy for having soils that are better drained than other communities dominated by northern white cedar.

Good Examples of Acadian Spruce - Fir Forest Natural Communities

HIGH-ELEVATION MOUNTAIN SLOPE FORESTS

High-elevation balsam fir forest: Mount Washington (Sargent's Purchase), Mount Carrigan (Lincoln), and Mount Bond (Lincoln).

High-elevation spruce - fir forest: Gibbs Brook Research Natural Area (Bean's Grant), Gore Mountain (Stratford), Great Gulf Wilderness Area (Thompson & Meserve's Purchase), Nancy Brook Research Natural Area (Livermore), The Bowl Research Natural Area (Waterville Valley), Carr Mountain (Warren/Wentworth) and Mount Kineo (Ellsworth).

LOWLAND VALLEY BOTTOM FORESTS

Lowland spruce - fir forest: Elbow Pond (Woodstock), Norton Pool (Pittsburg), Pondicherry Wildlife Refuge (Jefferson), South Bay Bog (Pittsburg), and Upper Ammonoosuc River (Kilkenny).

Montane black spruce - red spruce forest: Ethan Pond and Shoal Pond vicinities (Livermore).

Northern white cedar forest/woodland: Dearth Hill (Stewartstown).

Acadian Spruce - Fir Forest Wildlife

Arboreal mammals of Acadian spruce - fir forests include American marten and fisher. The marten is a state-threatened species common in areas with fallen logs. The logs provide cover for mice and voles, the marten's prey. The fisher, the American marten's larger cousin, inhabits dense conifer stands in the winter and preys on hares, rabbits, squirrels, mice, shrews, and porcupines. The rarest mammalian carnivore in New Hampshire, the Canada lynx, also makes its home in spruce - fir forests. This cat's large, furry feet allow it to stay on top of deep snow when it chases snowshoe hare. Moose winter in spruce - fir forests where they browse on buds, twigs, and needles of balsam fir, mountain ash, and yellow birch. Black bears use spruce - fir stands for escape, resting, or den sites when not seeking food in mixed deciduous woods.

Boreal chickadees nest and forage exclusively in spruce - fir forests, and spruce grouse survive on fir and spruce needles, a food source indigestible by other birds. Gray jays, red-breasted nuthatches, ruby-crowned kinglets, and dark-eyed juncos occur in high and low elevation spruce - fir forests. The milder weather at lower elevations favors northern parula and blue-headed vireos, species unable to withstand harsher conditions at significantly higher elevations.

Bicknell's thrush and blackpoll warbler breed exclusively in the higher elevation spruce - fir forest communities, especially in patches disturbed by storms and insect infestations. Globally, Bicknell's thrush is relatively secure, but it is a species of conservation concern because the state supports an estimated 40 percent of the bird's global breeding population. Blackpoll warblers make a grassy nest in a spruce tree to support their brood of four to five young. Other high-elevation bird residents include red- and white-winged crossbills, black-backed and three-toed woodpeckers, common raven, bay-breasted and blackburnian warblers, olive-sided flycatcher, and gray jay.

A boreal chickadee on Black Mountain in Benton.

A large black birch at Northwood Meadows State Park in Northwood.

Laurentian Mixed Forests

Laurentian mixed forests are hardwood and conifer forests of cool temperate climates. They extend from the Great Lakes to the greater St. Lawrence River region, covering much of northern New England and adjacent portions of eastern Canada. Laurentian refers to the St. Lawrence River and watershed that forms the heart of this continental forest region. The growing season is longer than in Acadian spruce - fir forests, and soils are more productive. These conditions favor a mixture of northern hardwoods and temperate climate conifers, and a diverse herbaceous layer.

In New Hampshire, Laurentian mixed forests consist of northern hardwood forests and hemlock - hardwood - pine forests. Northern hardwood forests cover many northern and higher-elevation parts of the state, and are the primary communities between 1,500 and 2,500 feet elevation. Northern hardwood species dominate, including sugar maple, beech, and yellow birch. Hardwoods mix with red spruce at higher elevations and hemlock at lower elevations. Northern wood sorrel, shining clubmoss, bluebead lily, twisted stalk, hobblebush, and striped maple are common understory species. In northern New Hampshire, northern hardwood forests occur adjacent to Acadian spruce - fir forests either above or below. In addition, they occur as small patches on mesic or slightly enriched sites below 1,500 feet elevation in central and southern parts of the state.

Hemlock - hardwood - pine forests occur farther south and at lower elevations than northern hardwood forests. They are most common below 1,500 feet elevation, and form a transition zone between northern hardwoods and Appalachian oak - pine forests. Hemlock - hardwood - pine forests are characterized by an increase in hemlock, white pine, and red oak, and a corresponding decline in sugar maple and yellow birch. Hemlock and beech are the dominant late successional species. Other common hardwoods include red maple and paper birch. Numerous sub-canopy species, such as witch hazel, maple-leaved viburnum, and wintergreen, are more abundant in these forests than in northern hardwood forests. Appalachian species such as white oak, black oak, hickories, and southern herbs and shrubs are sparse or absent.

The Laurentian mixed forest growing season is 120 to 150 frost-free days, compared to 90 to 120 days in Acadian spruce - fir forests. As a result, soils are more productive and nutrients are more available in Laurentian mixed forests. These conditions favor fast-growing, nutrient-demanding hardwoods over spruce and fir. Hardwoods concentrate more nutrients in their leaves than conifers. In the fall, nutrients not transferred back to stems and roots return to the soil when the leaves fall and decay. Consequently, hardwood forest surface humus contains more nutrients than soils of colder

Hobblebush is a common shrub in many Laurentian mixed forests.

climates, and acts like a slow-release fertilizer for the trees and plants that grow there.

Hardwoods do not dominate the entire Laurentian mixed forest. Temperate conifers, such as hemlock and white pine, out-compete hardwoods on poorer sites. Hemlock is particularly well adapted to rocky and shallow soils. White pine is most abundant on disturbed sites and sandy soils, including old fields once common in the region. Low-quality leaf litter of temperate conifer forests contributes to nutrient-poor soils and greater organic matter accumulation.

The distribution of natural communities and trees within the Laurentian mixed forest reflects regional gradients in elevation and latitude. For example, hemlock and red oak occur below 2,000 feet elevation, sugar maple and beech below 2,500 feet, and yellow birch below 3,000 feet. Northern hardwood forests dominate northern and higher-elevation parts of the region, but are patchily distributed at lower elevations and to the south.

The distribution and abundance of trees is also modified by regional differences in habitat availability and disturbance regimes. For example, cold-tolerant white pine might be expected to be more common in northern hardwood forests in the White Mountains, but infrequent fire, limited sand plain habitat, limited agricultural history, and susceptibility to blowdown limit its occurrence there.

Hurricanes and other major storms are primary large-scale disturbance agents in Laurentian mixed forests. They occur infrequently and affect most of the trees over large areas. Major storms can impact forests at any elevation or aspect. Historically, hurricanes have affected hemlock - hardwood - pine forests in southern and central parts of the state more than northern hardwood forests to the north. Smaller storms such as downbursts, tornados, nor'easters, and ice storms also cause substantial local damage.

Small-scale disturbances are more common. Chronic wind stress can be severe at high elevations in northern hardwood forests, and individual trees will fall periodically, creating small canopy gaps. Fungi and insects also cause much of the tree mortality, and predispose trees to wind damage by weakening stems, branches, and roots. Occasional fungi and insect outbreaks can also occur at a larger scale and affect whole forests. Overall, mammal and insect herbivores consume more biomass in Laurentian mixed forests than in Acadian spruce - fir forests, though the amount is still less than 10 percent of annual growth. Fires of natural origin are infrequent in Laurentian mixed forests, and only locally important on ridges and sand plains.

The formation of tree-canopy gaps modifies forest conditions and causes changes in tree species composition over time. Birches, aspen, and pin cherry are the primary gap colonizers in Laurentian mixed forests. Birches and aspen produce abundant lightweight seeds. These seeds are readily dispersed by wind, and germinate well on mineral soil associated with the root balls of fallen trees. Birch and aspen are shade-intolerant and require the elevated light, nutrient, and moisture levels found in gaps (moisture levels are higher because large trees no longer take up water in that location). Fast growth allows aspen and birch to out-compete shade-tolerant species. Pin cherry has a different strategy, relying on long-lived, buried seeds and good germination in large gaps. Birch, aspen, and pin cherry are short-lived, rarely exceeding 130 years of age.

In contrast to gap colonizers, shade-tolerant sugar maple, beech, hemlock, and red spruce require fewer nutrients and are longer-lived. Shade-tolerant species grow slowly in the understory, and ascend to canopy dominance following the demise of short-lived gap species. A few shade-tolerant trees can live for several hundred years. In New Hampshire, hemlock can exceed 500 years, while yellow birch and red spruce can exceed 400 years. Sugar maple and beech live for about 250 and 300 years, respectively.

Laurentian mixed forests feature a dense, diverse understory, comprised primarily of shade-tolerant shrubs and perennial herbs. Many of these species spread vegetatively, although some rely on seeds (or spores, on ferns) to persist. One growth strategy used by some herbaceous plants of these forests is to emerge and bloom before canopy leaf-out blocks sunlight. Such spring-flowering species include red and white trilliums, dwarf ginseng, Canada mayflower, trailing arbutus, and bellwort. By contrast, shade-intolerant species such as raspberries, blackberries, mountain maple, bindweeds, goldenrods, and bracken occupy gaps opened up by disturbance, and disperse their seeds by either wind or wildlife.

Pink lady's slippers are one of the most common orchids in Laurentian mixed forests.

Northern hardwood - spruce - fir forest at Mount Moosilauke in Benton.

Northern Hardwood Forests

Northern hardwood - spruce - fir forest occurs at intermediate elevations in cool, mesic settings in the mountains of central and northern New Hampshire. Substrates are shallow, rocky, nutrient-poor till or boulders. The community occurs between sugar maple - beech - yellow birch forests and either high- or low-elevation spruce - fir forests. These are mixed forests of red spruce, balsam fir, sugar maple, beech, and yellow, paper, and heartleaf birch. Sugar maple and beech disappear above 2,500 feet elevation, leaving only the birches, spruce, and fir. American mountain ash and Canadian honeysuckle occur in the shrub layer, and common herbaceous species include mountain wood fern, northern wood sorrel, and bluebead lily.

Sugar maple - beech - yellow birch forest is the most common hardwood forest type in northern New Hampshire, and is frequently referred to simply as northern hardwood forest. The community is most abundant on mountain slopes between 1,500 and 2,500 feet elevation, with patchier distribution at lower elevations. Sugar maple and beech are the dominant late-successional trees, and yellow birch is a long-lived gap colonizer. Spruce and fir are absent or sparse. Hobblebush is common in the shrub layer, while intermediate wood fern, shining clubmoss, bluebead lily, and Canada mayflower are common in the understory.

Sugar maple - beech - yellow birch forest at The Basin in Chatham.

Hemlock - spruce - northern hardwood forests occur on nutrient-poor soils in ravines, on river terraces, and on lower mountain slopes in the northern part of the state below 2,000 feet elevation. They are primarily conifer forests. Hemlock and red spruce are the dominant trees, with occasional balsam fir and white pine in some examples. Sugar maple, yellow birch, and beech are variable and usually sparse. The understory is similar to other northern hardwood forest types, with hobblebush, northern wood sorrel, and bluebead lily.

Hemlock - spruce - northern hardwood forest at Snyder Brook on Mount Madison.

Hemlock - oak - northern hardwood forest is a common mixed coniferous-deciduous forest of moderate elevations in central New Hampshire. The community is most abundant from 800 to 1,500 feet elevation, but it also occurs as patches on mesic sites in southern parts of the state, and in valley bottoms to 2,000 feet elevation. The canopy consists of classic northern hardwood species such as sugar maple, beech, and yellow birch mixed with hemlock. Red oak and white pine are also present, but become scarce in the mountains or at higher elevations. Striped maple, hobblebush, Indian cucumber root, partridgeberry, and intermediate wood fern are characteristic of the understory and ground cover. Soils are generally well drained, acidic, and nutrient-poor.

Hemlock - oak - northern hardwood forest at Gibbs Brook Scenic Area.

Hemlock - Hardwood - Pine Forests

Hemlock - beech - oak - pine forest is the most common upland forest community in southern and central New Hampshire below 1,500 feet elevation. The community represents a transition between northern hardwood forests and Appalachian oak and pine forests in terms of latitude, elevation, and vegetation. Hemlock and beech are the primary late-successional domi-

Hemlock - beech - oak - pine forest at Spalding Park Town Forest in Hollis.

Hemlock - white pine forest at College Woods in Durham.

Hemlock forest on North Mountain at Pawtuckaway State Park in Nottingham.

Beech forest on Mount Chocorua in Albany.

nants. Forests historically subjected to agriculture or timber harvesting are dominated by early to mid-successional species including red oak, white pine, red maple, and black or paper birch. Understory species vary widely, but commonly include witch hazel, wintergreen, and intermediate wood fern. Northern hardwood forest species, such as northern wood sorrel and bluebead lily, are absent or sparse.

Hemlock and white pine characterize **hemlock - white pine forests**, though sparse hardwoods are also possible in the canopy. The community is similar to hemlock - beech - oak - pine forest, but the understory is both sparser and more variable, and certain disturbance histories favor conifer dominance. The substrate is nutrient-poor rocky or sandy soils.

Hemlock forests are a common community of pure or nearly pure hemlock canopies. They feature dark, relatively open understories with few herbaceous species. Hemlock is shade-tolerant and maintains itself by outcompeting other tree species for light and nutrients. Hemlock forests serve as wintering habitat for deer and other wildlife species. Ravines and steep, rocky, north-facing slopes are classic locations for these forests.

Beech forests occur on coarse-till-covered hillsides in and south of the White Mountains. Beech dominates the canopy, sometimes with sparse red oak and red maple. Understory plants are few or absent, but may include beechdrops, a plant that parasitizes beech roots. The rare three-birds orchid also grows in litter-filled hollows in this community. Beech trees produce mast crops of nuts important to many animals, including black bear.

Dry red oak - white pine forests occur on dry sandy or rocky sites maintained by periodic fire. The community has an open forest canopy, and an understory dominated by lowbush blueberry, black huckleberry, bracken, and sweet fern. It is structurally and functionally similar to fire-maintained Appalachian oak and pine forest types, but lacks southern oaks and pitch pine. Red oak may dominate to the near exclusion of white pine when this community occurs on rocky ridges, whereas white pine is more abundant on sites with sandy soils. Some early successional hemlock - beech - oak - pine forests may superficially resemble this community. For example, red oak and white pine can dominate the overstory in forests that develop from abandoned pastures or in cutover areas on mesic or dry-mesic soils, but in the absence of drier soils and fire, hemlock and beech will eventually increase in prominence.

CHARACTERISTIC PLANTS OF SELECTED LAURENTIAN MIXED FOREST NATURAL COMMUNITIES

A = Northern hardwood - spruce - fir forest
B = Sugar maple - beech - yellow birch forest
C = Hemlock - spruce - northern hardwood forest
D = Hemlock - oak - northern hardwood for.
E = Hemlock - beech - oak - pine forest
F = Hemlock - white pine forest

COMMON NAME	SCIENTIFIC NAME	A	B	C	D	E	F
TREES							
Sugar maple	Acer saccharum	•	•	o	o	o	
American beech	Fagus grandifolia	•	•	o	•	•	
Yellow birch	Betula alleghaniensis	•	•	o	o	o	
Red spruce	Picea rubens	o		•	o		
Balsam fir	Abies balsamea	o			o		
Paper birch	Betula papyrifera	o	o	o		o	o
White ash	Fraxinus americana		o		o	o	
Hemlock	Tsuga canadensis			•	•	•	•
Red maple	Acer rubrum				o	o	
Red oak	Quercus rubra				o	•	o
White pine	Pinus strobus					•	•
Black cherry	Prunus serotina					o	
Black birch	Betula lenta					o	o
SHRUBS							
Mountain maple	Acer spicatum	o	o				
Hobblebush	Viburnum lantanoides	o	•	o	o		
Mountain ash	Sorbus spp.	o					
Canadian honeysuckle	Lonicera canadensis	o					
Pin cherry	Prunus pensylvanica		o		o	o	
Striped maple	Acer pensylvanicum		o	o			
Wintergreen	Gaultheria procumbens					o	o
Witch hazel	Hamamelis virginiana					o	o
Maple-leaved viburnum	Viburnum acerifolium						o
HERBS							
Intermediate wood fern	Dryopteris intermedia	•	•		o		o
Mountain wood fern	Dryopteris campyloptera	•	o	o			
Northern wood sorrel	Oxalis montana	o	o	o	o		
Bluebead lily	Clintonia borealis	o	o	o			
Shining clubmoss	Huperzia lucidula		o	o	o		
Canada mayflower	Maianthemum canadense		o				o
Starflower	Trientalis borealis		o	o			o
Painted trillium	Trillium undulatum		o	o			
Whorled aster	Oclemena acuminata	o	o				
Rose twisted stalk	Streptopus lanceolatus		o	o			
Wild sarsaparilla	Aralia nudicaulis		o	o		o	o
Indian cucumber root	Medeola virginiana					o	o
Partridgeberry	Mitchella repens					o	o
Goldthread	Coptis trifolia					o	
Sessile-leaved bellwort	Uvularia sessilifolia		o			o	o
Small whorled pogonia*	Isotria medeoloides*					o	

• = abundant to dominant   o = occasional or locally abundant   * = state threatened or endangered

Dry red oak - white pine forest on Bear Mountain in Lyme, with Smarts Mountain in the background.

Good Examples of Laurentian Mixed Forest Natural Communities

NORTHERN HARDWOOD FORESTS

Northern hardwood - spruce - fir forest: Lafayette Brook Scenic Area (Franconia), Greeley Ponds Scenic Area (Waterville Valley), Mount Sunapee State Park (Newbury), and above Gentian Pond (Success).

Sugar maple - beech - yellow birch forest: Mountain Pond (Chatham), The Bowl Research Natural Area (Waterville Valley), Mount Cardigan (Orange), Mount Sunapee State Park (Newbury), Nash Stream Forest (Stratford/Odell), Province Pond (Chatham), and Stoddard Rocks (Stoddard).

Hemlock - spruce - northern hardwood forest: Bartlett Experimental Forest (Bartlett), Dry Brook (Waterville Valley), McDonough Brook (Chatham), Swift River (Albany), and along the Peabody River (Gorham), and Wild Ammonoosuc River (Landaff).

Hemlock forest: Shingle Pond (Chatham), Pawtuckaway State Park (Nottingham), Pierce Reservation (Stoddard), and parts of Hemenway State Forest (Tamworth).

Beech forest: Chase Hill (Albany), Hammond Trail (Albany), and The Basin (Chatham).

Hemlock - oak - northern hardwood forest: Bartlett Experimental Forest (Bartlett) and Pisgah State Park (Winchester).

HEMLOCK - HARDWOOD - PINE FORESTS

Hemlock - beech - oak - pine forest: Five Finger Point (Sandwich), Hemenway State Forest (Tamworth), College Woods (Durham), Sheldrick Forest Preserve (Wilton), and Chase Hill (Albany).

Hemlock - white pine forest: Hemenway State Forest (Tamworth), College Woods (Durham), Pawtuckaway State Park (Nottingham), Sheldrick Forest Preserve (Wilton), Thatcher Memorial Forest (Hancock), and Heath Pond Bog Natural Area (Ossipee).

Dry red oak - white pine forest: Rattlesnake Mountain (Rumney), Pine River State Forest (Ossipee), Ames Mountain (Wentworth), Sugarloaf Mountain (Benton), and Dinsmore Mountain (Sandwich).

Laurentian Mixed Forest Wildlife

Laurentian mixed forest communities are widespread in New Hampshire, and their resident animals are commonly encountered. Moose and deer browse on young trees and the understory. Acorns and beechnuts provide black bears with much-needed fat reserves in preparation for winter. Black bears also feast on seasonal blueberry and dogwood fruits, and search rotting logs for protein-rich insects throughout the summer. Coyotes and bobcats compete for snowshoe hares, cottontails, squirrels, small rodents, and white-tailed deer. Rare species include blue-spotted salamander, black racer, New England cottontail, and northern goshawk.

As in other habitats, wildlife populations in Laurentian mixed forests are a function of complex predator-prey relationships. Fisher, Cooper's hawks, and northern myotis bats hunt for porcupine, blue jays, and insects, respectively. Common ravens nest on cliffs (see chapter 2), but hunt mammals, birds, and carrion in hardwood forests.

Many of the state's abundant songbirds breed in these forests, including black-capped chickadees, robins, ovenbirds, and downy woodpeckers. The diverse tree canopy and complex understory vegetation create structural layers that provide niches for different bird species. Veeries turn over leaves and bits of bark as they forage for insects and spiders on the forest floor. Canada warblers feed and sing from within shrub thickets at forest edges. Red-eyed vireos sing from high in the canopy.

Several bats live in trees in Laurentian mixed forests. Eastern pipistrelles and eastern red bats hide under leaves, with females gathering in communal maternity roosts in hollow trees. Red bats prefer the largest hardwood trees in the forest, where they may sleep and rest behind the flaking bark. Hoary bats roost in foliage and woodpecker holes.

Moose eat aquatic vegetation in summer and twigs and buds in winter. They are New Hampshire's largest mammal.

Appalachian oak and pine forest in southern New Hampshire.

Appalachian Oak and Pine Forests

Appalachian oak and pine forests primarily occur in warm-temperate climates of the central and Appalachian states, but they extend into southern and coastal New Hampshire and Maine. In these forests the climate is warmer and drier, elevations are lower, the growing season is longer, and fire was historically more frequent than in the Laurentian and Acadian forests to the north.

Plants with southern distributions—whose ranges are centered in Appalachian states farther south—distinguish Appalachian oak and pine forests from other New Hampshire forests. These species are largely absent from Laurentian forests. They include white oak, black oak, scarlet oak, hickories, sassafras, pitch pine, scrub oak, mountain laurel, flowering and round-leaved dogwoods, tick-trefoils, sweet goldenrod, false foxgloves, and wild indigo. Red oak and white pine are abundant in both Laurentian and Appalachian forests. Oaks are most abundant on dry or dry-mesic glacial till soils, and pines are most abundant on dry sand plain soils. On wetter, nutrient-rich sites, oaks mix with hickory, black birch, maple, and diverse herbs and shrubs. Appalachian forests have more locally rare plant species than Laurentian forests, in part because many temperate forest species reach the northeastern end of their geographic ranges in central New England.

In New Hampshire, Appalachian oak and pine forest communities mainly occur in southern parts of the state. They are most abundant in the Connecticut River Valley, Merrimack River Valley, and Coastal Plain regions. They occur below 1,000 feet elevation on dry to dry-mesic glacial till soils of ridges and slopes, and on sand plain features such as outwash, eskers, kame terraces, and other ice-contact deposits. Individual natural communities of

Mesic Appalachian oak - hickory forest near Beaver Brook in Hollis.

Appalachian and Laurentian forests can occur adjacent to one another or mix, especially in transition areas between the two forest types. Appalachian communities also occur as patches in central New Hampshire on south-facing ridges and sand plains with a history of frequent fire.

Climate exerts a strong influence on the soils and vegetation of Appalachian oak and pine forests. Compared to Laurentian mixed forests, the climate in Appalachian forests is warmer and drier, and the growing season is longer (averaging 150 or more frost-free days). On hardwood sites, the warmer climate enhances decomposition and mixing of organic matter in soil and, in moist hardwood areas, the formation of a relatively thin surface layer of organic humus. On poor sand plain soils, conifers reinforce poor site conditions by producing poor-quality leaf litter low in nitrogen and high in slow-to-decay lignin. Leaf litter accumulates until reduced by fire.

Important disturbance agents in Appalachian oak and pine forests include fungi, insects, storms, wind, and fire. The effects of most of these agents are similar to those in Laurentian mixed forests of central and southern New Hampshire, although fire is more important in Appalachian forests. Fire-return intervals are less than 200 years in oak forests and less than 50 to 100 years in pine woodlands. Native Americans used fire to enhance hunting opportunities near settlements. Later, European settlers increased fire incidence in the region as a result of agriculture, timber harvesting, and transportation activities. Oaks and pines are well adapted to warm, dry conditions conducive to fire, whereas northern hardwoods and hemlock are not. In the absence of fire, fire-intolerant species such as red maple, beech, and birch increase in Appalachian oak and pine forests.

The rare small whorled pogonia grows in some oak - pine forests in New Hampshire.

Many Appalachian oak and pine forest plants exhibit a variety of special adaptations to fire and drought. Young oak and pitch pine and understory heath shrubs sprout readily from roots or stumps, allowing them to respond rapidly to damage from burns. Fire stimulates pitch pine cones to release their seeds, which germinate best in mineral soils exposed by fires hot enough to burn off the duff layer. Mature oak and pine trees develop thick bark that insulates the interior of the tree from fire. Plants in dry and fire-prone Appalachian forests emphasize below-ground growth to maximize capture and retention of limited water and nutrient resources.

Appalachian oak and pine forest communities have either a forest or woodland canopy structure. Red oak is usually abundant and mixes with Appalachian oaks such as white oak, black oak, and scarlet oak. Shagbark and pignut hickories are occasional. Other common trees include red maple, paper birch, black birch, gray birch, and ironwood. Common shrubs include lowbush blueberry, hillside blueberry, dangleberry, black huckleberry, maple-leaved viburnum, sweet fern, and witch hazel. Velvet-leaf blueberry and sheep laurel are occasional. Wintergreen, Pennsylvania and woodland sedge, common hairgrass, rough-leaved rice grass, poverty oat-grass, bracken, whorled loosestrife, and pinweeds are common herbs. Appalachian oak and pine forests have an upper elevation limit of about 1,000 feet.

Dry Appalachian oak forest at Hickory Hill in Pelham.

Oak and Mixed Forests

Dry Appalachian oak forests occur on dry, nutrient-poor, and well to extremely well-drained soils of rocky hills and sand plains. Red, white, black, and scarlet oak dominate the tree canopy, and hickories are sometimes present. Heath shrubs are the most abundant group of species in the understory, sometimes forming a dense shrub layer. Herbs are relatively sparse. This community is at the northernmost extent of its range in south-central New Hampshire, where it is primarily restricted to south-facing slopes.

   Pitch pine - Appalachian oak - heath forest is an uncommon community of the lower Merrimack River Valley. The canopy has a forest or woodland structure and includes pitch pine, red oak, white oak, black oak, and scarlet oak. White pine is sometimes present. Lowbush and hillside blueberry, black huckleberry, and sheep laurel comprise a well-developed heath shrub layer. The herb layer is sparse, and can contain sweet goldenrod, round-headed bush-clover, and stiff-leaved aster.

Pitch pine - Appalachian oak - heath forest at Derryfield Park in Manchester.

   Chestnut oak dominates or co-dominates **chestnut oak forest/woodlands** on rocky ridges, hilltops, and hillsides. Canopy associates include red oak, white oak, and white pine. Heath shrubs are common in the understory. Chestnut oak is a dry-site species that reaches the northeast end of its range in New Hampshire and has a limited distribution in the state and

Chestnut oak distinguishes **chestnut oak forest/woodland** from other oak - pine forests.

Mesic Appalachian oak - hickory forest at Great Bay NWR in Newington.

region compared to other Appalachian oaks. The vegetation of this community closely resembles that of dry Appalachian oak forest.

Mesic Appalachian oak - hickory forests have a diverse canopy, including white, black, scarlet, and red oaks, shagbark hickory, white ash, white pine, hemlock, birches, maples, and beech. The shrub and herb layer is sparse to moderately developed, and commonly includes wild sarsaparilla, Canada mayflower, poison ivy, partridgeberry, and wintergreen. Low heaths and other dry-site understory plants are absent or sparse. The community occurs on fine-textured soils and other mesic or dry-mesic settings.

Oak - mountain laurel forest at Mount Wantastiquet State Forest in Hinsdale.

A dense and sometimes nearly impenetrable understory layer of mountain laurel distinguishes **oak - mountain laurel forests** from all other forest types. Broadleaf evergreen shrubs like mountain laurel are uncommon in upland forests of New Hampshire. The tree canopy is variable, and includes red and Appalachian oaks, hemlock, white pine, birches, and red maple. Other than mountain laurel, shrubs and herbs are sparse or absent. Oak - mountain laurel forest is restricted to south and southwest New Hampshire and parts of the Lakes Region.

Pitch pine - scrub oak woodland at West Branch Pine Barrens Preserve in Madison.

Pine Forests and Woodlands

Pitch pine - scrub oak woodlands, sometimes called simply pine barrens, occur in sand plain settings with extremely well-drained soil and a history of frequent fire. Pitch pine forms a woodland canopy and scrub oak dominates the understory. Common associates include lowbush blueberry, sweet fern, bracken, and distant sedge. Fire is required at least every 50 years to maintain the community. The absence of fire allows oak trees and fire-intolerant white pine, red maple, and birches to occupy canopy gaps. A large number of rare butterflies and moths, including the federally endangered Karner blue butterfly, occur in pine barrens. The largest and least-fragmented example of this community in New Hampshire occurs in the Ossipee region, though smaller, remnant patches occur in Concord and other locations in the Merrimack River Valley.

Mixed pine - red oak woodland at Heath Pond Bog Natural Area in Ossipee.

A mix of red and white pine, with little or no pitch pine, characterizes **red pine - white pine forests**. Balsam fir is sometimes prominent in the understory, and mosses are common. The fire-return interval required to maintain this community is 70 or more years, longer than the 50-year return interval for pitch pine - scrub oak woodlands. The community occurs on moderately dry to moist sand plains of central New Hampshire, near the southern limit of red pine. Uncommon in New Hampshire, fire-maintained red and white pine forests are abundant in the northern Great Lakes region.

An unusual mix of pitch, red, and white pines, along with red oak, characterizes **mixed pine - red oak woodlands**. These woodlands occur on eskers, outwash, and other sand plain settings with fire-return intervals of at least 70 to 100 years. Scrub oak is occasional but not dominant in the understory. Fire-intolerant species such as gray birch, big-toothed aspen, and red maple are common. Mixed pine - red oak woodlands are restricted to central New Hampshire where the distributions of pitch and red pine overlap.

Good Examples of Appalachian Oak and Pine Natural Communities

OAK AND MIXED FORESTS

Dry Appalachian oak forest: Jeremy Hill State Forest (Pelham) and Pawtuckaway State Park (Nottingham).

Pitch pine - Appalachian oak - heath forest: Derryfield Park (Manchester) and Ponemah Plain (Amherst).

Chestnut oak forest/woodland: Beaver Brook Association land (Hollis), Dumplingtown Hill (Raymond), and Pawtuckaway State Park (Nottingham).

Mesic Appalachian oak - hickory forest: Mount Wantastiquet (Hinsdale), Great Bay National Wildlife Refuge (Newington), and Pawtuckaway State Park (Nottingham).

Oak - mountain laurel forest: Mount Wantastiquet (Hinsdale), Sheldrick Forest (Wilton), Beaver Brook Association land (Hollis), Chase Hill (Albany), and along the margins of Squam Lake (Holderness).

PINE FORESTS AND WOODLANDS

Pitch pine - scrub oak woodland: Concord Pine Barrens (Concord) and Ossipee Pine Barrens (Madison, Ossipee, Tamworth).

Red pine - white pine - forest: Pine River State Forest (Ossipee) and in the Big River drainage (Barnstead/Strafford).

Mixed pine - red oak woodland: Pine River State Forest (Ossipee and Effingham), White Lake State Park (Ossipee), south of Cedar Swamp Pond (Kingston), and the Moat Brook vicinity (Hale's Location).

CHARACTERISTIC PLANTS OF SELECTED APPALACHIAN OAK AND PINE FOREST NATURAL COMMUNITIES

A = Pitch pine - scrub oak woodland
B = Dry Appalachian oak forest
C = Red pine - white pine forest
D = Mixed pine - red oak woodland
E = Mesic Appalachian oak - hickory forest
F = Oak - mountain laurel forest

COMMON NAME	SCIENTIFIC NAME	A	B	C	D	E	F
TREES							
Red oak	Quercus rubra		•	o	•	•	o
Black oak	Quercus velutina		•			•	
White oak	Quercus alba		•			o	o
Scarlet oak	Quercus coccinea		•				
Chestnut oak	Quercus montana		•				o
White pine	Pinus strobus	o	o	•	•	•	
Red pine	Pinus resinosa			•	•		
Pitch pine	Pinus rigida	•	o		•		
Balsam fir	Abies balsamea			•			
Shagbark hickory	Carya ovata		o			o	o
Pignut hickory	Carya glabra		o				
Red maple	Acer rubrum		o		o	•	o
Paper birch	Betula papyrifera			o		o	o
Black birch	Betula lenta		o			•	o
Sassafras	Sassafras albidum		o				o
White ash	Fraxinus americana					o	o
American beech	Fagus grandifolia					o	o
Ironwood	Ostrya virginiana					o	
Hemlock	Tsuga canadensis		o			o	o
SHRUBS							
Scrub oak	Quercus ilicifolia	•	o		o		
Beaked hazelnut	Corylus cornuta	o				o	o
Poison ivy	Toxicodendron radicans					o	
Lowbush blueberry	Vaccinium angustifolium	o	o	o	o	o	
Black huckleberry	Gaylussacia baccata	o	o			o	
Sweet fern	Comptonia peregrina	o			o		
Maple-leaved viburnum	Viburnum acerifolium					o	o
Flowering dogwood	Cornus florida		o				o
Sheep laurel	Kalmia angustifolia		o		o		
Mountain laurel	Kalmia latifolia		o			o	•
Striped maple	Acer pensylvanicum						o
HERBS							
Rough-leaved rice grass	Oryzopsis asperifolia	o		o			o
Pennsylvanian sedge	Carex pensylvanica		o			o	o
Sweet goldenrod*	Solidago odora*		o				
Rattlesnake weed	Hieracium venosum		o				
Wild sarsaparilla	Aralia nudicaulis					o	
Bracken	Pteridium aquilinum	o		o			o
Canada mayflower	Maianthemum canadense			o		o	
Indian cucumber root	Medeola virginiana						o

• = abundant to dominant   o = occasional or locally abundant   * = state threatened or endangered

Wild lupine is a state-rare plant.

Federally endangered Karner blue butterfly.

Pine Barrens and Fire

Pitch pine - scrub oak woodlands are commonly referred to as pine barrens. The community is rare in New Hampshire, and supports several species of rare plants and songbirds whose populations are in decline. In addition, a large number of rare butterflies and moths inhabit pine barrens, and are dependent on the community's specialized plants. For example, the federally endangered Karner blue butterfly requires wild lupine to complete its life cycle, and in New Hampshire wild lupine is abundant only in frequently burned areas. An area must burn at least once every 50 years to maintain pine barrens habitat.

Most of the plant species in pine barrens are adapted to fire. Pitch pine, an early successional species, is an excellent example. Fire clears the forest floor of competing vegetation, exposes the mineral soil, and increases available light. Some pitch pine cones are serotinous, opening only after they are exposed to extreme heat. The seeds are released when forest duff has burned, exposing mineral soil needed for germination. The thick bark of mature pitch pine trees allows them to withstand light burning. New branches are produced by growing needle clumps directly on the trunk in a process called epicormic sprouting. Pitch pine can also produce stump sprouts after the main stem has been killed, creating a thick whorl of new growth around the stump.

In New Hampshire, the absence of fire in pine barrens allows oak trees and fire-intolerant white pine, red maple, and birches to occupy canopy gaps. To maintain healthy pine barrens habitat, land managers sometimes conduct prescribed fires. Fires are carefully controlled and only implemented under safe conditions. In any given year, only small portions of the community are burned to allow other areas time to recover. Regular fires also protect neighboring properties by reducing the woody fuel loads that accumulate over time, reducing the risk of uncontrolled wildfires.

(Left) One year after a prescribed burn in the West Branch Pine Barrens Preserve.

(Right) Trail through a healthy pine barrens community several years after a controlled burn.

Appalachian Oak and Pine Forests Wildlife

Appalachian oak and pine forests comprise less than 10 percent of New Hampshire's total land area, yet they support a disproportionately large number of vertebrate wildlife species, including eight amphibians, twelve reptiles, sixty-seven birds, and seventeen mammals. This high species richness likely results from the large amount of niche diversity in these forests.

The sandy glacial outwash and wetlands occurring throughout these forests provide the essential mix of habitat for many reptiles. Common turtles such as painted and snapping turtles, and even the rare Blanding's turtle, may be seen crossing roads as they travel from wetlands to dry, sandy areas to nest and lay eggs during the summer months. Eastern hognose snakes, a state-threatened species, also lay eggs in the sandy, well-drained soils, and feed on amphibians assoicated with wetlands of Appalachian oak and pine forest wetlands.

Vernal pools in these forests provide vital breeding habitat for amphibian species such as spotted salamanders, toads, and wood frogs. The temporary nature of these pools precludes the establishment of fish populations that would consume eggs and developing tadpoles.

Oaks and pines are mast species, which means they produce large crops of seeds, or mast, on a cyclical basis. Mammals feed on the mast, but cannot consume all of the seeds produced in strong mast years, ensuring that some seeds will germinate and mature into trees. This cyclical mast production reduces a tree's resource expenditure in low or non-mast years.

The distinctive nasal "peent" of American woodcock and the namesake call of the whip-poor-will are two sounds of spring in Appalachian oak and pine forests. Both of these bird species are at risk because of decreasing acreage of early successional habitat associated with these forests. Historically, wildfires and other natural disturbances maintained diverse age classes of vegetation across the forested landscape. Fire suppression, combined with habitat loss from development, has reduced structural complexity in some of these forests, resulting in fewer habitat niches and wildlife species.

Maidenhair fern in rich woods in the Connecticut River Valley.

Rich Woods

Rich woods are floristically diverse, nutrient-enriched hardwood forest communities. They are associated with particular combinations of bedrock, soils, topography, and moisture. Sugar maple, white ash, and a species-rich herbaceous layer are hallmarks of rich woods. Shrubs are sparse, but ferns, perennial forbs, and sedges are abundant, including many species that flower in early spring. Rich woods are uncommon in New Hampshire, and provide habitat for the majority of the state's rare forest plant species.

Rich woods typically occur as patches in deciduous forests of eastern North America, including the Laurentian and Appalachian forests of New Hampshire. They are closely related to mixed mesophytic forests of the central Appalachians, one of the most diverse temperate forest regions of the world, and an important glacial refugium for many of New Hampshire's forest plants. The similarity in species composition among these disparate patches of rich woods suggests that certain site factors, particularly nutrient levels, have a greater influence on the distribution of rich woods communities than climate.

New Hampshire's rich woods are divided into rich mesic, rich dry-mesic, and semi-rich forests. Rich mesic forests are very moist and nutrient-enriched, and contain the most diverse and lush herbaceous layer of any New Hampshire forest type. Characteristic species include maidenhair and ostrich ferns, blue cohosh, wood nettle, northern waterleaf, Dutchman's breeches, and plantain-leaved sedge. Rich dry-mesic forests occur on steep, south-facing slopes, and contain species indicative of rich, warm, dry, and rocky conditions. Characteristic trees include sugar maple, red oak, ironwood, and occasionally hickory. The herbaceous layer is diverse, but shorter and less well-developed than rich mesic forests. Herbs present include early saxifrage, rock cresses, rusty woodsia, ebony spleenwort, blue-stemmed goldenrod, blackseed mountain rice, Pennsylvania sedge, and herb Robert. Semi-rich forests are less enriched and diverse than rich mesic and rich dry-mesic forests. They contain species indicative of moderately enriched conditions, such as Jack-in-the-pulpit, trilliums, baneberries, foamflower, red elderberry, Christmas fern, and zigzag goldenrod.

Calcium and nitrogen are the key plant nutrients of rich woods. Nutrient levels are a function of soil and bedrock characteristics, water flow and availability, and topographic setting. Calcium is stored in organic matter and on exchange sites on soil mineral particles. Calcium helps sustain relatively low soil-acidity levels, neutralizes harmful levels of soluble minerals like manganese and aluminum, improves soil structure, enhances microbial activity, and increases the availability of other nutrients. Nitrogen, derived from the atmosphere and stored in organic matter, affects the growth rate and productivity of forest plants. Microbes and other soil organisms are the engines that drive rapid cycling of nutrients between soils and plants, as they actively mix the forest soil and transform nitrogen and other nutrients bound in organic compounds into forms available to plants.

Moisture levels and water flow affect biological activity and the delivery of nutrients to the root zone. Plant growth and soil microbe activity are enhanced when water is continuously available, and reduced during drought or inundation. Moisture levels in rich woods vary from very moist year-round in small seepage zones to being characterized by dry soil. Xeric conditions occur only on a very small scale within rich woods, such as on bare outcrops or boulders.

Nutrient, moisture, and solar energy levels are affected by topographic setting. Rich woods commonly occur in colluvial topographic settings such as coves, benches, narrow valleys, and the bases of slopes. These sites are convergence zones for the downslope movement of nutrient-bearing leaf litter, sediments, and water. Slope steepness and aspect affect the amount of solar energy a site receives. For example, steep, south-facing slopes receive considerably more solar energy than steep north-facing slopes. More solar energy means higher temperatures, a longer growing season, lower moisture levels, and increases in organic matter decomposition and site productivity. Most rich woods occur in colluvial settings on hills with southern aspects, and rich dry-mesic forests occur almost exclusively on steep, rocky, south-facing slopes.

Plants of rich woods are well adapted to the nutrient-rich environment. Many require soil conditions with relatively low acidity and higher calcium levels. Conversely, they are intolerant of the more acidic, nutrient-poor, high-aluminum environment of other forest soils. The high nutrient supply in rich woods also favors plants with the ability to grow fast and acquire nutrients rapidly over those with the ability to use fewer nutrients more efficiently. Rich woods plants allocate the majority of their resources to stem and leaf growth, and lesser amounts to their roots. Consequently, many rich woods plants have high total leaf area, enhancing their ability to capture light in the shady understory. In addition, rich woods plants concentrate nitrogen in their foliage, which enables high rates of photosynthesis and biomass production. Much of the nitrogen in the foliage of deciduous trees and herbs is translocated to stems and roots prior to leaf fall. Translocation notwithstanding, nitrogen and calcium concentrations are higher in rich-site leaf litter than poor sites characterized by beech, oaks, and conifers. Rich-site litter also contains fewer carbon-rich compounds (such as lignin) that slow microbial decay.

Spring ephemerals are a specialized group of perennial forest herbs largely restricted to rich woods. Examples of spring ephemeral plants in-

Trout lily and spring beauty, two rich-site indicator plants, in bloom in a rich forest at Coleman State Park in Stewartstown.

Squirrel corn, a spring ephemeral wildflower species, in bloom at Yatsevitch Forest in Plainfield.

clude Carolina spring beauty, squirrel corn, Dutchman's breeches, wild leeks, and trout lilies. These plants leaf and flower in early to mid spring, before hardwood leaves fully emerge. Above-ground portions of spring ephemerals diminish or disappear after fruiting. This strategy allows spring ephemerals to reduce competition for light and nutrients, although they expend considerable energy resources in the process. Herbivores are attracted to the high nitrogen content of spring ephemeral leaves, though the short time that the plants spend above ground limits their losses to herbivores. More susceptible to herbivory are spring-flowering plants with leaves that persist through the summer, such as blue cohosh, wild ginger, hepaticas, and ginseng.

Rich mesic forest at Yatsevitch Forest in Plainfield.

Wakerobin in flower in **rich red oak rocky woods** in Sanbornton.

Rich Mesic Forests

Rich mesic forests are the most diverse hardwood forests in New Hampshire. The soils are moist and enriched with nutrients. Sugar maple is the dominant tree, and white ash and basswood are less abundant associates. Ferns, forbs, and wide-leaved sedges are abundant. Characteristic herbs include maidenhair fern, silvery spleenwort, blue cohosh, northern waterleaf, American ginseng, Goldie's fern, sweet cicely, squirrel corn, plantain-leaved sedge, and bloodroot. Herbaceous spring ephemerals include Dutchman's breeches, trout lily, broad-leaved toothwort, and wild leek. Examples of this community on high floodplains, steep river terraces, and till hillsides in the Connecticut River Valley support bitternut hickory, bladdernut, and showy orchis, species that approach the northeastern limit of their geographic range in New Hampshire.

Rich Dry-Mesic Forests

Rich red oak rocky woods have a woodland structure and occur on steep, rocky, south-facing slopes below 2,000 feet from the White Mountains south. The thin canopy is dominated by red oak and sugar maple, with occasional white ash, basswood, and ironwood. The understory flora is diverse and contains rich-site species such as hepatica, flat-leaved sedge, early saxifrage, rusty woodsia, blue-stemmed goldenrod, large yellow lady's slipper, black-seed mountain rice, Pennsylvania sedge, and herb Robert. Other common plant species include wild columbine, marginal wood fern, false Solomon's seal, pussytoes, and poison ivy. The rare fern-leaved false foxglove and rock sandwort may grow on ledges or outcrops. Compared to rich mesic forests, this community has a more open canopy, the herb layer is less dense, and the species composition reflects rockier and drier conditions.

Rich Appalachian oak rocky woods are restricted to elevations below 1,000 feet and distinguished by Appalachian species that reach their northern limit in southern New Hampshire. These species include white oak, hickories, flowering dogwood, ebony spleenwort, Venus' looking-glass, and many rare plants including Missouri and smooth rock cresses, skydrop aster, downy false foxglove, hairy stargrass, early buttercup, and hoary mountain mint. In addition, many plants of rich red oak rocky woods also occur in this community.

Red oak - ironwood - Pennsylvania sedge woodland is an uncommon natural community of upper slopes and ridges in southern and central New Hampshire (see also chapter 2). Oaks, hickories, white ash, ironwood, and sugar maple form an open, park-like canopy over lawns of Pennsylvania sedge. Herbaceous plants indicative of rich, dry conditions are usually pres-

The state-rare early buttercup in **rich Appalachian oak rocky woods** at Pawtuckaway State Park in Nottingham.

ent, including hepatica, rusty woodsia, ebony spleenwort, blue-stemmed goldenrod, and sometimes rare species such as rock cresses and reflexed sedge. Other herbs common to rocky woods are found here as well, such as wood anemone, wild columbine, and marginal wood fern.

Semi-Rich Forests

Semi-rich mesic sugar maple forests occur throughout the state up to 2,600 feet elevation. Sugar maple is the dominant tree, white ash is common,

Semi-rich mesic sugar maple forest at Lafayette Brook Scenic Area in Franconia.

Semi-rich oak - sugar maple **forest** at Cave Mountain in Bartlett.

and beech and yellow birch are occasional. The community is less enriched than rich mesic or rich dry-mesic forests, but more enriched than forests that dominate most of New Hampshire. Species diversity and productivity are lower than rich mesic forests because of lower moisture or soil nutrient levels, or both. Consequently, semi-rich forests harbor indicators of weakly enriched conditions such as Jack-in-the-pulpit, baneberries, Christmas fern, trilliums, foamflower, red elderberry, and zigzag goldenrod.

Semi-rich oak - sugar maple forest occurs in central and southern New Hampshire, mostly below 1,500 feet. The community is similar to semi-rich sugar maple forests, but contains red oak, ironwood, hickories, and other southern or drier-site species. Appalachian species such as black oak and shagbark hickory may be present in southern New Hampshire below 1,000 feet elevation. Semi-rich indicators include Christmas fern, red baneberry, blue-stemmed goldenrod, and flat-leaved sedge. Indicators of strong enrichment, such as blue cohosh, wild ginger, maidenhair fern, and Goldie's fern, are absent. Common hardwood forest plants that are frequently present include partridgeberry, intermediate wood fern, and sessile-leaved bellwort. Semi-rich oak - sugar maple forests on steep or flat river terraces have diverse tree and shrub layers.

Good Examples of Rich Woods Natural Communities

Rich mesic forest: Sugarloaf Cove and Sundown Ledge (Albany), Cape Horn (Northumberland), Pawtuckaway State Park (Nottingham), Yatsevitch Forest (Cornish/Plainfield), Coleman State Park (Stewartstown), and Weeks State Park (Lancaster).

Rich red oak rocky woods: Rattlesnake Mountain (Rumney), Bald Knob (Moultonborough), Whites Ledge (Bartlett), and Devil's Slide (Stark).

Rich Appalachian oak rocky woods: Jeremy Hill (Pelham), Merrill Hill (Hudson), Gilmore Hill (Merrimack), Pawtuckaway State Park (Nottingham), Crommet Creek (Newmarket), and Mount Wantastiquet (Hinsdale).

Red oak - ironwood - Pennsylvania sedge woodland: West Rattlesnake Mountain (Holderness), Daniels Mountain (Hinsdale), Warwick Preserve (Westmoreland), and Pawtuckaway State Park (Nottingham).

Semi-rich mesic sugar maple forest: Mountain Pond Research Natural Area (Chatham) and Sugarloaf Mountain (Haverhill)

Semi-rich oak - sugar maple forest: Crommet Creek vicinity (Durham), Merrimack River Conservation Center (Concord), and Pawtuckaway State Park (Nottingham).

CHARACTERISTIC PLANTS OF SELECTED RICH WOODS NATURAL COMMUNITIES

A = Rich red oak rocky woods
B = Semi-rich oak - sugar maple forest
C = Semi-rich mesic sugar maple forest
D = Rich mesic forest

COMMON NAME	SCIENTIFIC NAME	A	B	C	D
TREES					
Sugar maple	Acer saccharum	●	●	●	●
White ash	Fraxinus americana	o	●	o	o
Basswood	Tilia americana	o	o	o	o
Butternut	Juglans cinerea	o			
Black cherry	Prunus serotina	o	o	o	
Red oak	Quercus rubra	●	●	o	
White oak	Quercus alba	o	o		
Shagbark hickory	Carya ovata	o	●		
Bitternut hickory	Carya cordiformis	o			
American elm	Ulmus americana	o			
Red maple	Acer rubrum	o		o	
Ironwood	Ostrya virginiana	o	●	o	o
Yellow birch	Betula alleghaniensis	o		●	o
American beech	Fagus grandifolia			o	o
SHRUBS					
Maple-leaved viburnum	Viburnum acerifolium	o	●		
Witch hazel	Hamamelis virginiana	o	o		
Flowering dogwood	Cornus florida		o		
Musclewood	Carpinus caroliniana		o		
Beaked hazelnut	Corylus cornuta		o	o	
Alternate-leaved dogwood	Cornus alternifolia			o	
Red elderberry	Sambucus racemosa			o	
HERBS					
Trout lily	Erythronium americanum	o			
False Solomon's seal	Maianthemum racemosum	o			
Blunt-lobed hepatica	Anemone americana	o	o		
Blue cohosh	Caulophyllum thalictroides	o			o
Wild ginger	Asarum canadense	o			o
Zigzag goldenrod	Solidago flexicaulis			o	
Broad beech fern	Phegopteris hexagonoptera		o		
Christmas fern	Polystichum acrostichoides		o	o	
Round-leaved violet	Viola rotundifolia		o	o	
Baneberries	Actaea spp.		o	o	
Wide-leaved sedges	Carex spp.		o		o
Silvery spleenwort	Deparia acrostichoides			o	o
Maidenhair fern	Adiantum pedatum				o
Ostrich fern	Matteuccia struthiopteris				o
Wood nettle	Laportea canadensis				o
Mountain sweet cicely*	Osmorhiza berteroi*				o
Squirrel corn	Dicentra canadensis				o
Goldie's fern	Dryopteris goldiana				o

● = abundant to dominant   o = occasional or locally abundant   * = state threatened or endangered species

Rich Woods Wildlife

Some common mammals that use rich woods include gray squirrel, eastern chipmunk, white-tailed deer, and black bear. Common birds include wood thrush, veery, ovenbird, black-and-white warbler, Louisiana woodthrush, red-eyed vireo, pileated woodpecker, and barred owl. With the exception of several insects dependent on spring ephemeral plants, few animals are specifically limited to rich woods. However, the spring flush of nutritious and palatable rich wood plants attracts herbivores at a time when few other food sources are available.

The high levels of calcium in rich woods affect wildlife in many of New Hampshire's forests. One example of this influence concerns a fairly inconspicuous animal: the land snail. Land snails are an important link between environmental calcium and animals higher on the food chain. Snail shells are primarily composed of calcium carbonate, which the snails obtain by consuming plant materials, fungi, animal waste, nematodes, and other snails. They also absorb calcium from small rocks, outcrops, and soil. Other animals—including beetles, turtles, salamanders, shrews, mice, squirrels, thrushes, ruffed grouse, and wild turkeys—obtain calcium by eating land snails. The calcium they obtain from the shells is important for many vital functions, such as building bones and manufacturing eggshells. Consequently, changes in land snail populations can lead to substantial impacts to other animals. For example, acid precipitation reduces soil calcium, resulting in fewer snails, which in turn causes eggshell thinning and reduced reproductive success in birds.

Many herbs of rich woods utilize an interesting dispersal mechanism called myrmecochory, in which seeds are moved by ants. For example, bloodroot—one of the earliest spring wildflowers in areas with mineral-enriched soils—has a small drop of fatty oil on the exterior of its seed. Ants collect the seeds and return them to their colonies. There they consume the oil, which provides a valuable source of energy, then discard the seeds in a waste chamber within the colony. This is beneficial for the seed, as it now has a nutrient-rich site in which to germinate, safely below the soil surface and out of the reach of seed-predators like birds and rodents. Other plants that use this dispersal strategy include hepatica, wild ginger, Dutchman's breeches, and trout lily. The strategy does have drawbacks, however. Because ants do not generally travel great distances, the plant's dispersal distance is equally limited. This becomes particularly problematic in areas where forests are fragmented by agricultural fields or development. If one of these species disappears from an isolated forest patch, its limited dispersal ability virtually ensures that it will not return, absent human intervention.

4 Peatlands

Peatlands, also known as bogs, fens, mires, muskegs, and moors, are saturated wetlands containing layers of partially decayed plant remains called peat. Walking across a soggy peatland surface can be a peculiar experience, as the spongy peat mats sometimes bounce and quake underfoot. A fascinating variety of plants are uniquely adapted to the waterlogged, oxygen-deprived conditions of the peatland environment, and many of these plants are found in no other habitat. The dominant life-forms are peat mosses, sedges, forbs, dwarf to tall shrubs, and trees; the relative abundances of these organisms, both living and dead, vary greatly among peatland types. This chapter primarily considers open peatlands, those with less than 25 percent cover of trees. Peatlands with more than 25 percent tree cover are considered along with other forested wetlands in chapter 5.

Peatlands are classified as either bogs or fens, the definitions of which have been the subject of debate among scientists. Some apply the term "bog" only to raised peatlands that are entirely rain-fed and completely isolated from runoff. Such raised peatlands do not occur in New Hampshire. Others apply the term more loosely to any acidic peatland containing peat mosses. In this book, bog refers only to extremely acidic, heath shrub dominated peatlands with vegetation similar to raised, rain-fed bogs. All other open peatlands in New Hampshire are considered fens.

In fens, mineral input from water or the land surrounding a peatland is a stronger influence than in bogs. As a result, fens have a lower abundance of heath shrubs and a greater abundance of sedges and non-heath shrubs. Nutrient levels distinguish three types of fens: poor, medium, and rich. Poor fens most closely resemble bogs. They are highly acidic, have limited nutrient input, and frequently occur adjacent to heath shrub bogs. Medium and rich fens are less acidic to alkaline, more enriched, and better oxygenated than bogs or poor fens. They rarely occur adjacent to heath shrub bogs, and have a greater diversity of sedges, mosses, herbs, and shrubs than poor fens or bogs.

Peatlands develop in saturated areas where the growth rate of plants exceeds the decomposition rate of their remains. Over time, this imbalance results in the build-up of thick deposits of peat. Peatlands are most abundant in cold, wet, or humid parts of the northern hemisphere, such as the boreal forest and sub-arctic regions, where precipitation nearly equals or exceeds the loss of water to the atmosphere by evaporation. Peatlands are also lo-

PRECEDING PAGE: Arethusa, a rare peatland plant, at Copps Pond in Tuftonboro.

BELOW: Bog and poor fen communities ring Little Cherry Pond (*top*) and medium fen communities flank the margins of its inlet stream (*bottom*).

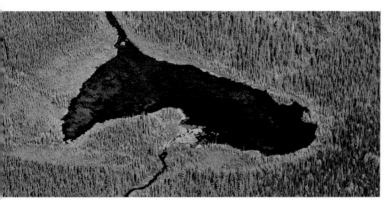

cated in New England and other temperate climate regions where locally saturated conditions occur in wetland basins with very poor or slow drainage. In these settings, plant growth exceeds decomposition because soils are permanently saturated, acidic, and poorly aerated. These factors impede the activity of the microbes that drive decay, mainly fungi and bacteria. Peat mosses (the genus *Sphagnum*) and anaerobic bacteria directly affect decomposition and peat build-up. Peat mosses slow decomposition by increasing the acidity of their environment, whereas anaerobic bacteria, which thrive less than a foot below the permanent water level, produce compounds such as organic acids and alcohol that inhibit decomposition.

Peatlands form by paludification or lake-fill. Paludification is the process in which peatlands creep onto and transform upland sites. This phenomenon occurs in response to a rising local water table, due to changes in land use, shifts in regional climate, or when peat growth and accumulation impedes drainage. Paludification is far more important than lake-fill in northern and temperate-maritime regions. In most parts of New Hampshire, however, the climate is neither cold enough nor wet enough for extensive paludification to occur. Settings where paludification does occur in New Hampshire include alpine ridges and cliff brows, broad outwash plains with a high water table, and along the upland margins of lake-fill peatlands.

Lake-fill peatland formation occurs when lakes and ponds fill over thousands of years with organic matter. This is the most common way peatlands form in temperate climate regions, including in most of New Hampshire. The general successional progression in these settings is from open water to fen to bog to conifer swamp. These changes occur in large measure because the plants and their remains modify the hydrology and chemistry of the environment as peat accumulates. In northern New Hampshire, a slightly different succession may occur in lake basins fed by calcium-rich waters. The environments in these lakes have a greater abundance of aquatic life, including algae, other plants, and small invertebrates, than non-enriched lake basins. The difference in initial conditions produces a progression from floating and grounded rich fen to northern white cedar swamp.

The degree of infilling evident in contemporary peatlands, and the extent to which vegetation zones develop uniformly within a basin, depends on basin depth, the steepness of the basin's slopes, and nutrient input. The classic lake-fill sequence begins with the establishment of pioneering plants along the water's edge. These include some combination of sedges, mosses, aquatic herbs, and leatherleaf (a shrub), all of which grow well in the nutrient-poor conditions found in many of New Hampshire's ponds and small lakes. Initially, these plants create a soupy mass along the shore. Over time, this loose aggregation slowly becomes firmer and thicker, eventually

MEDIUM AND RICH FEN SETTINGS

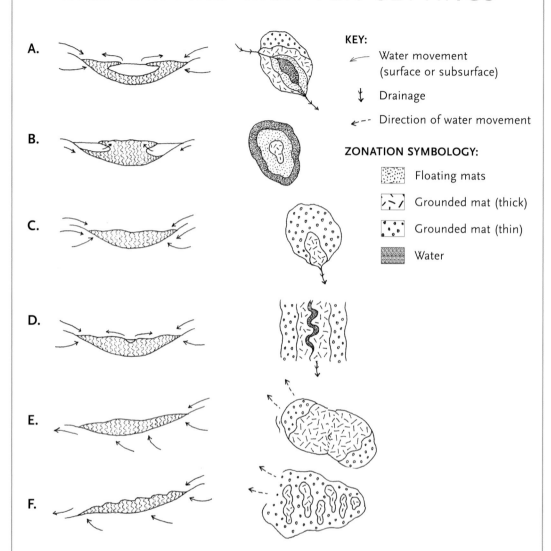

KEY:

$\swarrow$ Water movement (surface or subsurface)

$\updownarrow$ Drainage

$\twoheadleftarrow$ Direction of water movement

ZONATION SYMBOLOGY:

Floating mats

Grounded mat (thick)

Grounded mat (thin)

Water

A. Partially filled, open lake basins (stream inlets and outlets) with floating and grounded mats

B. Basins with prominent moat (lagg) along upland border

C. Open basins, with grounded mats affected by combinations of surface and near-surface runoff and groundwater flow

D. Open basins and low-gradient stream channels affected by combinations of surface and near-surface runoff, groundwater flow, and streambank overflow

E. Groundwater-fed slopes in cold climate areas of northern New Hampshire (not patterned)

F. Groundwater-fed slopes in cold climate areas of northern New Hampshire (patterned)

forming a floating sedge mat at the lake's edge. The mat also provides organic debris that settles and accumulates on the lake bottom, causing the lake to fill from both the bottom upward and the top downward. Aeration decreases as the peat mat thickens, impeding decomposition and fostering peat moss growth. The mosses increase the acidity of their environment, which further slows decomposition. As the peat mat continues to develop, zones of vegetation form around the lake, often irregularly, but sometimes creating a concentric pattern. The composition of these zones depends on whether the mat is floating or grounded, and on the distribution of nutrients across the peatland. Portions of grounded mats farthest from open water and the margin of the upland are slightly higher, drier, and less influenced by mineral-enriched water than floating mats and upland border areas. Heath shrubs, and eventually black spruce, are favored on grounded mats, where hummock-and-hollow microtopography forms and contrasts with the flatter surface of floating mats.

Regional or local changes in water level shift the dominant peatland vegetation. A rise in water level promotes continued peat build-up, whereas a drop in water level accelerates peat decomposition. In the New England region, historic climate changes induced fluctuations in water levels and corresponding shifts in peatland vegetation. The cool climate that initially followed glacial retreat produced saturated conditions that favored peatland expansion. This trend lasted until about 7,600 years ago, when a shift to a warmer, drier climate caused regional water tables to drop. Peat decomposition increased in response to the warmer climate, and peatlands shifted to more wooded conditions. The climate then cooled about 2,500 to 3,000 years ago, again raising regional water levels, drowning wooded peatlands and returning the communities to open bogs.

Local water-level changes, whether natural or anthropogenic, have similar effects as climate change, only these effects are accelerated. Beavers construct dams across the outlet streams of peatlands, flooding and killing trees and shifting vegetation dominance to shrubs or herbaceous plants. Tree removal by timber harvesting, disease, or fire also leads to higher water levels, as water formerly transpired by the vegetation remains in the soil instead.

Nutrient availability influences peatland vegetation, and nutrients are less available in most peatland types than in other kinds of wetlands. Few nutrients are accessible to plants in strongly acidic bogs; nutrient levels increase in less acidic to alkaline fens. Many nutrients are bound up in the accumulated peat and thus unavailable to plants. Peatland plants must therefore be efficient at procuring and retaining nutrients, especially nitrogen and phosphorous, the availability of which limits the rate of plant growth and overall biological productivity. Most peatland plants possess physiological tolerances limited to very specific ranges of mineral concentrations and pH.

BOG AND POOR FEN SETTINGS

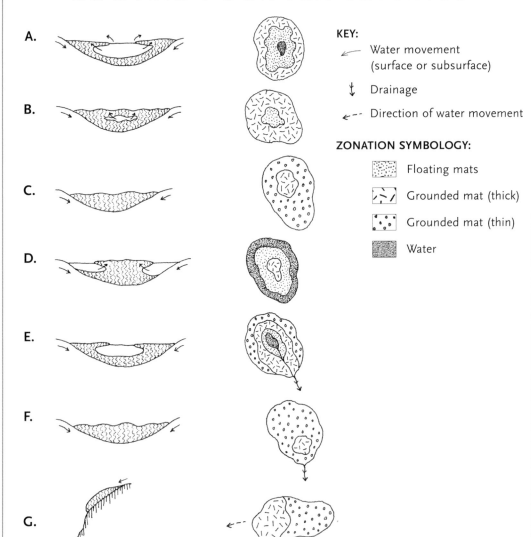

KEY:

⬿ Water movement (surface or subsurface)

↕ Drainage

⬸-- Direction of water movement

ZONATION SYMBOLOGY:

▦ Floating mats

▨ Grounded mat (thick)

▨ Grounded mat (thin)

▨ Water

A. Partially filled, closed lake basin (no stream drainage) with floating and grounded mats

B. Filled, closed lake basin with floating mat over remnant lens of lake water and surrounding grounded mat

C. Filled, closed lake basin with grounded mat

D. Basin with prominent moat (lagg) along upland border

E. Partially filled, open lake basin (stream drainage) with floating and grounded mat

F. Filled, open lake basin with grounded mat

G. Peat mat on sloped terrain in cold climates at high elevation, including cliff brows, margins of ridgeline depressions, and deep snowpack zones

As a result, any shift in nutrient levels or pH can produce a shift in species composition.

Calcium, magnesium, iron, and aluminum all influence peatland species composition, in part due to mineral toxicity. High levels of aluminum and iron in bogs and poor fens are toxic to the roots of many plants. As a result, bog and poor fen peatlands contain only species tolerant of high aluminum and iron levels. Conversely, the high calcium levels of rich fens are toxic to many bog and poor fen plants. Rich fens only contain species that tolerate or require elevated calcium levels.

Leatherleaf occurs in many peatland communities.

Some plants have growth forms and life-history characteristics that enable them to contend with the deficiencies or high concentrations of certain minerals in peatlands. For example, heath shrubs, which are especially abundant in bogs and poor fens, efficiently secure and conserve both nitrogen and phosphorous, while tolerating high levels of aluminum. Heath shrubs maximize below-ground root growth at the expense of stem growth, a strategy that focuses resources on obtaining limited soil nutrients. Several heath shrub species have evergreen, compact, waxy, and crowded leaves that both conserve nutrients and reduce the need for stem growth. In addition, heath shrub roots take up nutrients in acidic, high-aluminum conditions that cripple the roots of other plants. Conversely, the high calcium levels in rich fens inhibit heath shrubs, as well as most peat mosses. Finally, carnivorism is an unusual approach to overcoming the nutrient-deprived conditions of bogs and fens. Sundews, pitcher plants, and bladderworts trap and digest insects and other invertebrates, an advantage over plants that rely solely on soil nutrients.

Yellow sedge primarily occurs in rich fens.

At least 550 plant species grow in New Hampshire's peatlands, including more than 400 vascular and 100 bryophyte species. This amounts to more than 20 percent of the species in each of those plant groups within the state. Heath shrubs and peat mosses dominate bogs, while a more diverse mix of plants occurs in fens, including non-heath shrubs, sedges, and other herbaceous plants. Certain plant families are heavily concentrated in open or wooded peatlands, including more than 30 orchids, 80 sedges, and nearly all of the state's 43 known species of peat mosses. Many peatland plants are northern, cold-climate species that approach their southern distributional limit in New Hampshire, though a few are southern species that reach their northern limit here. Half of the 83 rare plant species occurring in New Hampshire's peatlands occur in rich fens, which occupy only a very small proportion of the total peatland area in the state. These plants do not tolerate or compete well under the more acidic and calcium-poor conditions found in other peatlands. The high species diversity in New Hampshire's peatlands reflects the state's ecological crossroads location and the presence of an extremely broad range of climate, elevation, and geologic conditions.

PEATLANDS AT A GLANCE

	BOGS	POOR FENS	MEDIUM FENS	RICH FENS
Dominant Vegetation	Dwarf heath shrubs	Moss or moss-sedge carpets	Sedges and medium height shrubs	Sedges and forbs
Mosses and Liverworts	Peat mosses dominant		Peat mosses abundant	Peat mosses sparse; other mosses and liverworts abundant
Shrubs	Heath shrubs abundant, dwarfed (<20 in. tall): leatherleaf, sheep laurel		Non-heath shrubs abundant, taller (>20 in. tall): sweet gale, meadowsweet	Shrubs variable, tall; willows, dogwoods present
Sedges	Few sedges; sparse, short		Sedges abundant and taller, including wire, bottle-shaped, and tussock sedges	Sedges diverse and abundant, including inland, yellow, and porcupine sedges
Forbs and other Herbs	Absent or rare		Common in low abundance including marsh St. John's-wort, swamp candles, cinnamon fern	High diversity and abundance, including orchids and other forbs, ferns, rushes, and grasses
Perimeter Trees and Tall Shrubs	Black spruce, Atlantic white cedar, mountain holly, and highbush blueberry		Red maple, larch, winterberry, and alder	Northern white cedar, alder, willow, and dogwood
Acidity – Alkalinity	Extremely acidic: pH from 3s to low 4s		Acidic: pH from mid-4s to mid-5s	Weakly acidic to weakly alkaline: pH from 6 to 8
Setting	Kettle holes, other closed basins, and open basins with very stagnant drainage		Open basins, with outlets and better drainage	Groundwater seepage zones influenced by calcium-rich bedrock or till

Peat Moss

Peat mosses (those in the genus *Sphagnum*) are giants in both size and influence within the diminutive world of mosses and liverworts. They overshadow all other mosses, with stems often an inch wide and a foot or more in length. Their long, central stems and numerous branches bear dozens of triangular, overlapping leaves that wick water to ever-growing tips. This adaptation is especially important because peat moss lacks vascular tissue. Peat mosses can hold ten to twenty-five times their weight in water, one of the reasons they are very well adapted to wet environments.

Peat mosses perpetuate their own surroundings to a greater extent than most plants. They acidify soil water by releasing hydrogen ions. At the same time, they absorb calcium and other minerals from the water, retaining those nutrients in their leaves and stems and rendering them unavailable to other plants.

Peat mosses are sensitive indicators of their surroundings. Each species has a particular preference or tolerance for light, nutrients, and water. Some grow as soupy masses in pools, whereas others occur only on drier, sunny hummock tops or in cool, shady, conifer swamps. Fifteen to twenty Sphagnum species commonly occur in a single New Hampshire peatland, indicating subtle variations in ecological conditions. Most peat moss species grow in acidic peatlands, but a few grow only in calcium-rich fens or upland settings such as wet cliffs or gravelly shores. Peat mosses also exhibit inter- and intraspecific variation in color, size, and growth habit in response to varying conditions, adding much to a peatland's visual diversity.

Forty-three of the sixty-one species of peat mosses in North America occur in New Hampshire. This species richness is higher than in many northern regions of similar size, reflecting the great range of climatic, geologic, and environmental conditions within the state. Most are boreal or alpine species, but several hug the Atlantic coastal plain and reach their northern range limit in or near New Hampshire.

Peatlands occupy a relatively small percentage of New Hampshire's landscape. In Alaska, Canada, northern Europe, and Russia, however, extensive peatlands are harvested for fuel and agricultural amendment. Since peat soils store a substantial proportion of the world's carbon, these mosses also play an important role in regulating global climate.

Sphagnum moss dominates bogs and poor fens.

Bogs and Poor Fens

Bogs and poor fens are the quiet haunts of peat mosses, heath shrubs, black spruce, and carnivorous plants. The scene is remarkably similar in many of these peatlands: a still, black-water pond often forms a "bog eye" at the center of the basin, surrounded by a springy carpet of peat mosses peppered with carnivorous plants such as sundew and pitcher plant. Farther from the water, heath shrubs dominate where the mat becomes firmer. Finally, spires of black spruce ring the outer perimeter of the bog, resembling the spruce and heath muskegs of boreal North America. All plants decompose very slowly in the wet, acidic conditions of bogs and poor fens, leading to the accumulation rather than the diminishment of their organic remains.

Bogs and poor fens occur in saturated and very poorly drained settings such as level basins with no or limited outflows. These settings remain saturated year-round. The only nutrient inputs are from precipitation, scant runoff from a small watershed, and sometimes pond or lake water. Kettle holes are the classic geologic setting for bogs in New England. These basins formed wherever giant chunks of relict glacial ice were buried in glacial outwash or other coarse ice-contact deposits. The ice chunks melted, leaving behind depressions with no inlets or outlets. Some depressions intersect the water table, forming mineral-deprived ponds in time. These conditions are perfect for the establishment of bogs. Bogs and poor fens also occur on broad, flat outwash plains with high water tables, or on slopes and in concavities on ridges at high elevation. Sloping bogs are rare in New England, only occurring high in the mountains where abundant precipitation, cloud-intercept, late-melting snowbanks, and a short growing season maintain perennially wet conditions.

Natural community distribution in bogs and poor fens corresponds to minor variations in nutrient and water levels. Peatlands with a central pond have a floating mat that rings the pond and supports poor fen communities. The saturated, floating mat is relatively flat and has a continuous carpet of peat moss and a relatively sparse cover of carnivorous plants, short sedges, and creeping or dwarfed mats of small cranberry and other heaths. Farther from the pond edge the peat mat is firmer and thicker, and densely covered by somewhat taller but still relatively short heath shrubs, including leatherleaf, sheep and bog laurels, and sometimes huckleberries. The shrubs grow over a continuous, undulating carpet of peat mosses, with drier hummocks and wetter hollows. Black spruce appears in this part of the peatland, and increases in abundance and height toward the upland edge, sometimes transitioning to a black spruce swamp. A wet, soupy transition area, called a moat or lagg, often occurs along the upland edge of the peatland. Herbaceous or shrub vegetation more akin to marshes or swamps occupies the

Star Lake on Mount Madison supports an **alpine/subalpine bog**.

moat, where increased inputs of nutrients from upland runoff and seasonal fluctuations of the water table accelerate decomposition of organic matter. Peatlands without a central pond lack a floating mat, or the floating mat is located over the deepest part of the basin and obscures a lens of water below the surface.

Species composition in bogs and poor fens varies throughout the state. High-elevation bogs (above 3,500 feet) have arctic-alpine species that are both disjunct from their centers of distributions far to the north, and absent from the state's low-elevation peatlands. Alpine bilberry, black crowberry, cloudberry, and Lindberg's peat moss are examples of such geographically disjunct arctic-alpine species. Bogs and poor fens south of the White Mountains contain numerous plants absent from more northern peatlands. Species with this southern distribution include highbush blueberry, huckleberry, male berry, and Virginia chain fern. Peatlands within 30 miles of the New Hampshire coast contain coastal plain specialists such as dwarf huckleberry, sweet pepperbush, and several rare or uncommon peat mosses not found farther inland.

Bog and poor fen communities include moss and moss–sedge carpets, dwarf heath shrub bogs, and tall shrub poor fens. Herbaceous marsh communities that sometimes occur around the margins of these peatlands are described in the medium and rich fens section. In contrast to medium and rich

(Left) Pitcher plant at Mud Pond Bog in Hillsborough.

(Right) Grass-pink is a common peatland orchid.

fen communities, all bog and poor fen communities are extremely to very acidic (pH from 3 to low 4s) and lack plants indicative of mineral-enriched conditions. Common plants of bogs and poor fens include leatherleaf, sheep and bog laurel, small and large cranberry, pitcher plant, three-seeded sedge, and tawny cotton-grass.

Moss and Moss–Sedge Carpets

Poor fen moss carpets have a relatively flat surface and support vegetation of low stature. Hummocks are generally less than 6 inches tall. They share an abundance of bryophytes and a paucity of dwarfed heath shrubs. Peat mosses, liverworts, and occasionally other mosses form extensive carpets on the surface of floating or loosely consolidated peat mats, in wet pool areas within firmer mats, or in weakly mineral-enriched moats adjacent to uplands. Heath shrubs generally contribute less than 25 percent cover and are less than 1.5 feet in height. Short sedges range from scattered to moderately abundant. The peat is very acidic, and poorly decomposed near the surface. Moss and moss–sedge carpets occur most often in peatlands with dwarf heath shrub bog communities, but some can occur in peatlands with medium fen communities. Two communities, **bog rosemary - sedge fen** and **subalpine sloping fen**, are intermediate in plant composition and structure between moss–sedge carpets and dwarf heath shrub bogs.

Sphagnum rubellum - **small cranberry moss carpets** occur on floating peat mats, and sometimes on adjacent grounded mats, in kettle holes and other isolated peatland basins. The reddish *Sphagnum rubellum* forms

Sphagnum rubellum - **small cranberry moss carpet** at Philbrick-Cricenti Bog in New London.

a nearly continuous carpet with lesser amounts of other peat mosses. The diminutive small cranberry is also a diagnostic species, and leatherleaf is always present. Other occasional plants growing on the carpet include bog and sheep laurels, bog rosemary, pitcher plant, tawny cotton-grass, hare's tail, and white beak-rush. Trees are absent or sparse and stunted. This widespread community is one of the most acidic peatlands in the state, often with a pH more acidic than vinegar (pH 3.9 or less).

The **liverwort - horned bladderwort mud-bottom** community forms low, floating mats in flat depressions. Dense, turf-like carpets known as mud-bottoms are comprised of a tiny, leafy liverwort that turns black and resembles mud from a distance. Various acid-tolerant peat mosses are common, as are spatulate-leaved sundew, pitcher plant, white beak-rush, small cranberry, bog rosemary, and leatherleaf. Most examples of this community

Liverwort - horned bladderwort mud-bottom at Little Church Pond in the White Mountains.

Large cranberry - short sedge moss lawn in Pittsburg.

type occur in kettle holes as small patches within other floating mat communities, adjacent to pools, or in other interior, very acidic areas isolated from upland runoff. It is concentrated in southern and central New Hampshire where kettle holes are most abundant. The community is visually striking when the yellow flowers of horned bladderwort are in bloom.

The **large cranberry - short sedge moss lawn** community is widespread in New Hampshire, forming on floating mats or in moats adjacent to uplands. Aquatic peat mosses dominate the loosely consolidated carpets, with lesser amounts of short sedges and scattered shrubs. Large cranberry, spatulate-leaved sundew, and short sedges such as three-way sedge, white beak-rush, and silvery sedge are common. These species are more frequent and abundant than in the previous two communities, indicating slightly higher nutrient levels. Pitcher plant, tawny cotton-grass, and other common bog plants may also be present.

Bog rosemary - sedge fen south of Ossipee Lake.

Bog rosemary - sedge fen occurs throughout the state on grounded mats or thick floating mats. Various peat mosses dominate the mat, and leatherleaf, bog rosemary, bog laurel, few-seeded sedge, bottle-shaped sedge, and three-leaved false Solomon's seal are common. Sweet gale is sometimes present. Most examples are extremely acidic. The abundance of shrubs and sedges varies among examples. Some occurrences approach a dwarf heath shrub structure, but the relatively flat surface and the presence of poor fen sedges align the community with the moss–sedge carpet types.

The **montane level fen/bog** community occurs in small bedrock depressions in exposed montane settings (mostly above 2,500 feet). It lacks both the arctic-alpine species found in high-elevation bogs (above 3,500 feet) and many of the species with a more southern distribution that occur in lower elevation peatlands of south-central New Hampshire. Organic soils

Montane level fen/bog on Mount Monadnock.

Subalpine sloping fen above Cannon Cliff.

range from thin veneers over bedrock to deeper peat mats. Species composition and structure vary depending on water level, nutrient status, soil depth, microtopography, and other factors. Common species may include peat mosses, cotton grasses, beak rushes, and other sedges, round-leaved sundew, three-leaved false Solomon's seal, cinnamon fern, leatherleaf, rhodora, mountain holly, and other shrubs. Red spruce and other tree species are sometimes sparsely present at the edge of these mountain peatlands.

Subalpine sloping fens are steeply sloped peat mats located on the brows of some high-elevation cliffs. Their structure can range from moss–sedge carpets to dwarf heath shrub bogs. This community is similar to alpine/subalpine bog, but has pioneer peat moss species and abundant herbs including mountain avens and Pickering's reed bentgrass. When heavily saturated, these peat mats are subject to sloughing, sometimes dislodging the entire community when they slide over cliff edges.

Dwarf Heath Shrub Bogs

Dwarf heath shrub bog communities have vegetation of low to moderate stature, and are distinguished from moss and moss–sedge carpet types by a denser cover of dwarf heath shrubs. Total shrub cover generally exceeds 25 percent, and the shrubs are usually dwarfed (less than 1.5 feet tall) or of medium height (up to about 2.5 feet tall). Tall shrubs are sparse or absent. Black spruce or larch may be present, but with a lower stature and abundance compared to those occurring in wooded fens. Hummocks are larger and taller (greater than 6 inches in height) compared to moss and moss–sedge carpet communities. Surface peat is poorly to moderately well decomposed, and very to extremely acidic (pH 4.1 or lower). Leatherleaf - sheep

Leatherleaf - sheep laurel shrub bog at Spruce Swamp in Fremont.

laurel shrub bogs and leatherleaf - black spruce bogs are level, and located in lowland and montane settings below 2,900 feet elevation. Alpine/subalpine bogs and wooded subalpine bog/heath snowbanks are either level or sloping, and restricted to higher elevations.

Leatherleaf - sheep laurel shrub bogs are widespread, have an abundance of dwarf- to medium-height heath shrubs such as leatherleaf, sheep laurel, and rhodora, and have few tall shrubs and trees. Overall species richness is low. Mossy hummock and hollow topography is well developed, with occasional sedges on the hummocks. Some examples are extremely acidic (pH in the mid 3s) and have a very low shrub layer less than 6 inches in height. Others are less acidic (pH in the low 4s) and have somewhat taller shrubs, ranging from 1.5 to 2.5 feet in height.

Leatherleaf - black spruce bogs resemble leatherleaf - sheep laurel shrub bogs. Stunted black spruce and larch trees are present more often in this community and have a slightly higher cover. They also resemble muskeg habitats of the boreal forest region in more northern climates. The community lacks tall shrubs, but dwarf heath species such as leatherleaf, Labrador tea, sheep laurel, bog laurel, and small cranberry are abundant. The peat mats are grounded, and peat moss hummocks are usually well developed. The community is widely distributed in the state.

Leatherleaf - black spruce bog at Watts Wildlife Sanctuary in Effingham.

Alpine/subalpine bogs form in level depressions and adjacent slopes on high-elevation (above 3,500 feet) mountain ridges. They contain plants indicative of saturated conditions, such as small cranberry, hare's tail, and tussock bulrush, and an abundance of peat moss. The community partly resembles wooded subalpine bog/heath snowbanks in that it supports alpine species absent in montane and lowland bogs, such as alpine bilberry, black crowberry, and cloudberry.

Wooded subalpine bog/heath snowbanks occur on level or sloping terrain where deep, late-melting snowpacks contribute to saturated conditions and a short growing season, and in drier settings adjacent to alpine/subalpine bogs. The wet-site species of alpine/subalpine bogs are absent, but thick peat soils support abundant dwarfed heath shrubs, black spruce, and balsam fir.

Tall Shrub Poor Fens

Tall and medium-height shrubs with scattered trees dominate these wooded fens. In contrast to dwarf heath shrub bogs, tall shrubs comprise at least 15 percent cover and often form dense thickets. Trees form a sparse but distinct overstory. Both tall shrub poor fen communities are more acidic and nutrient-poor than the tall shrub fens associated with medium and rich fens.

Highbush blueberry - mountain holly wooded fen occurs mostly as a border thicket around more-open dwarf heath peatlands in southern and central parts of the state. A mixture of tall and medium-height shrub species is common, including highbush blueberry, mountain holly, leatherleaf, sheep laurel, and black huckleberry. Herbaceous plants are relatively sparse, but may include three-seeded sedge, Virginia chain fern, and three-leaved false Solomon's seal growing among abundant peat moss in the hollows.

Mountain holly - black spruce wooded fen occurs in northern New Hampshire where it forms border thickets around dwarf heath shrub bogs in lake-fill peatlands, or occupies extensive areas in broad peatland basins. Mountain holly, witherod, and shorter dwarf heath shrubs are the most abundant species, with variable amounts of black spruce and larch forming a scattered tree layer. The community resembles highbush blueberry - mountain holly wooded fen, but lacks southern species such as highbush blueberry, male berry, black huckleberry, and Virginia chain fern.

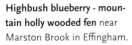

Highbush blueberry - mountain holly wooded fen near Marston Brook in Effingham.

Good Examples of Bog and Poor Fen Natural Communities

MOSS AND MOSS–SEDGE CARPETS

Sphagnum rubellum - **small cranberry moss carpet**: South Bay Bog (Clarksville), Lost Ponds (Ossipee), Red Hill Pond (Sandwich), Cedar Swamp Pond (Kingston), Trask Swamp (Alton), and Little Church Pond (Livermore/Albany).

Liverwort - horned bladderwort mud-bottom: South Bay Bog (Clarksville), Lost Ponds (Ossipee), Cedar Swamp Pond (Kingston), Little Church Pond (Livermore/Albany), and White Lake Kettles (Tamworth).

Large cranberry - short sedge moss lawn: Lake Umbagog (Errol), Cedar Swamp Pond (Kingston), and Little Church Pond (Livermore/Albany).

Bog rosemary - sedge fen: Between Ossipee Lake and Route 25 (Ossipee), at the delta of the Bearcamp River (Ossipee), and in the Broad/Leavitt Bay kettles (Ossipee).

Montane level fen/bog: Mount Cardigan (Orange), Mount Kearsarge (Warner), and Mount Monadnock (Jaffrey).

Subalpine sloping fen: Top of Cannon Cliff in the White Mountains.

DWARF HEATH SHRUB BOGS

Leatherleaf - sheep laurel shrub bog: East of Pine River (Ossipee), Bradford Bog (Bradford), Rochester Heath Bog (Rochester), and Big Church Pond (Livermore).

Leatherleaf - black spruce bog: Northwest of Umbagog Lake (Errol), Whaleback Ponds (Errol), South Bay Bog (Clarksville), Trask Swamp (Alton), Hubbard Pond (Rindge), and Loverens Mill (Antrim).

Alpine/subalpine bog: Bald Cap, Mount Success, Shelburne-Moriah Mountain, Mount Jackson, Mount Eisenhower, and Mount Adams.

Wooded subalpine bog/heath snowbank: Imp Mountain, Eagle Crag, Mount Success, Mount Moriah, Mount Hight, Mount Jackson, and Mount Eisenhower.

TALL SHRUB POOR FENS

Highbush blueberry - mountain holly wooded fen: South of Ossipee Lake (Ossipee), east of Pine River (Effingham), northwest of Umbagog Lake (Errol), Spruce Swamp (Fremont), and Big Church Pond (Albany).

Mountain holly - black spruce wooded fen: Sweat Meadows (Errol), Pontook Reservoir (Dummer), Whitewall Mountain (Bethlehem), and Ossipee Mountains (Ossipee).

CHARACTERISTIC SPECIES OF SELECTED BOG AND POOR FEN NATURAL COMMUNITIES

A = *Sphagnum rubellum* - small cranberry moss carpet
B = Liverwort - horned bladderwort mud-bottom
C = Large cranberry - short sedge moss lawn
D = Leatherleaf - sheep laurel shrub bog
E = Leatherleaf - black spruce bog
F = Highbush blueberry - mountain holly wooded fen

COMMON NAME	SCIENTIFIC NAME	A	B	C	D	E	F
SHRUBS & SAPLINGS							
Bog rosemary	*Andromeda polifolia*	o	o				
Sweet gale	*Myrica gale*			●			
Small cranberry	*Vaccinium oxycoccos*	●	o			●	
Leatherleaf	*Chamaedaphne calyculata*	●	o	●	●	●	●
Sheep laurel	*Kalmia angustifolia*	o			●	●	●
Bog laurel	*Kalmia polifolia*	o				●	o
Large cranberry	*Vaccinium macrocarpon*			●			o
Rhodora	*Rhododendron canadense*				o		o
Black spruce	*Picea mariana*					●	o
Larch	*Larix laricina*					●	o
Highbush blueberry	*Vaccinium corymbosum*						●
Mountain holly	*Nemopanthus mucronatus*						●
HERBS							
Pitcherplant	*Sarracenia purpurea*	o	o	o			
White beak-rush	*Rhynchospora alba*	o	●	●			
Horned bladderwort	*Utricularia cornuta*		●				
Round-leaved sundew	*Drosera rotundifolia*		o				
Spatulate-leaved sundew	*Drosera intermedia*		●	●			
Tawny cotton-grass	*Eriophorum virginicum*	o		o		o	
Hare's-tail	*Eriophorum vaginatum*	o				o	
Three-way sedge	*Dulichium arundinaceum*			●			
Silvery sedge	*Carex canescens*			●			
Swamp candles	*Lysimachia terrestris*			o			
Marsh St. John's wort	*Triadenum virginicum*			o			
Billing's sedge	*Carex billingsii*				o	o	o
NON-VASCULAR							
Liverwort	*Cladopodiella fluitans*		●				
Peat moss	*Sphagnum cuspidatum*		●	●			
Peat moss	*Sphagnum rubellum*	●	o		o	●	o
Peat moss	*Sphagnum torreyanum*			●			
Peat moss	*Sphagnum pulchrum*			●			
Peat moss	*Sphagnum capillifolium*				●	o	o
Peat moss	*Sphagnum magellanicum*				o	●	●
Peat moss	*Sphagnum angustifolium*					●	o

● = abundant to dominant   o = occasional or locally abundant   * = state threatened or endangered

BOG AND POOR FEN
NATURAL COMMUNITIES

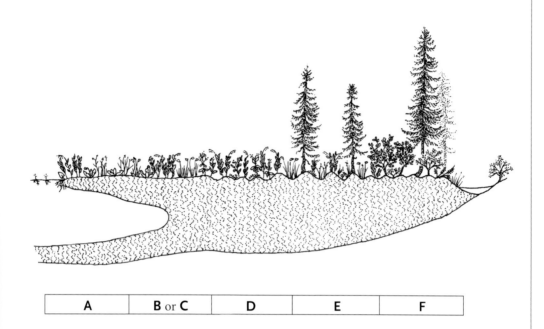

A	B or C	D	E	F

A.   **Liverwort - horned bladderwort mud-bottom** – Floating liverwort mats with short herbs and shrubs

B.   *Sphagnum rubellum* - **small cranberry moss carpet** – Floating mats of crimson red peat moss with carnivorous plants

C.   **Bog rosemary - sedge moss lawn** – Floating or soft peat moss lawns with sedges and dwarf shrubs

D.   **Leatherleaf - sheep laurel shrub bog** – Grounded hummocky mats with dwarf- to medium-height heath shrubs

E.   **Leatherleaf - black spruce bog** – Grounded, hummocky mats with black spruce

F.   **Highbush blueberry - mountain holly wooded fen** – Grounded, hummocky mats with tall shrubs and sparse trees or saplings

An idealized bog and poor fen natural community sequence. Other community combinations are possible.

A medium fen along the Merrymeeting River in Alton.

Medium and Rich Fens

Medium and rich fens are mineral and oxygen-enriched peatlands, with an overall character somewhere between bog and marsh. Sedges, forbs, and non-heath shrubs are more abundant in medium and rich fens than in bogs and poor fens, but the heaths and peat mosses prominent in the latter still occur. The wet-to-dry sequence in medium and rich fens is decidedly more herbaceous and deciduous than in bogs and poor fens. Tall sedges and non-heath shrubs occupy the wettest areas, and give way to winterberry, alder, herbs, and scattered larch and red maple toward the upland.

Nutrient status and species composition distinguish medium and rich fen natural community types. Medium fens are intermediate in nutrient status and acidity between bogs and poor fens at one extreme and rich fens at the other. Commonly associated with level basin settings, medium fens are moderately to very acidic, with a pH ranging from the mid 4s to mid 5s. Rich fens are less common than medium fens, are weakly acidic to weakly alkaline, and have higher levels of calcium and certain other minerals. They are associated with sloping seepage zones in regions of calcium-rich bedrock in northern New Hampshire. Medium and rich fens have many species in common, but each group contains species not found in the other. Many species found in rich fens are absent from all other types of peatlands.

The elevated nutrient and oxygen levels of medium and rich fens are products of runoff, seasonal water-level fluctuations, and groundwater seepage. Runoff carries minerals from surrounding landscapes into the fens. Seasonal drops in water level aerate the rooting zone of plants and accelerate decomposition of organic matter, increasing the yield of available nutrients. The most common settings for medium fens are along the borders of ponds, lakes, and sluggish streams. In contrast, fens on slopes or in level headwater basins that lack an internal pond or stream have dampened seasonal fluctuations and depend on upland runoff and seepage for minerals.

Wire sedge dominates a fen south of Ossipee Lake.

Sweet gale is a common shrub found in medium fens.

Medium and rich fens contain an abundance of sedges such as bottle-shaped sedge.

Vegetation patterns in level fens reflect differences in nutrient status and water levels within the peatland. Natural communities form concentric circles around a pond or lake, or parallel to a stream. Sedge fens occur in the wettest and most nutrient-enriched zones closest to ponds, streams, or upland borders, and are the most common communities on floating mats along pond margins. Characteristic plants include large cranberry, swamp candles, and numerous, robust sedges such as wire sedge, few-seeded sedge, bottle-shaped sedge, silvery sedge, and tussock sedge. Shrub or shrub–sedge dominated zones form on hummocky, grounded mats that are drier, but seasonally flooded. Sedges found in wetter areas mix with heath and non-heath shrubs, including leatherleaf, sweet gale, and meadowsweet. Taller wooded fens, with an abundance of tall and medium-height shrubs and scattered trees, occur toward the outer margin of the peatland. Winterberry, speckled alder, cinnamon fern, red maple, and larch are common plants of wooded fens. Peat mosses are less abundant in medium and rich fens than in poor fens and bogs.

Most medium level fens found in pond basins or along drainages are subject to damming and flooding as a result of beaver or human activity. Inundation of a grounded fen mat typically causes a shift to emergent marsh vegetation. In contrast, floating mats rise with the water, and vegetation may change little, if at all. Marsh plants can persist in these flooded basins until the water level drops following dam abandonment or the basin slowly fills with peat, changes that favor a return to fen vegetation.

Several unusual medium and rich fen types occur in the northern part of the state. Sloping seepage fens lack internal ponds or streams and exhibit different vegetation patterns than level fens. Many of these fens are relatively small and have weak internal zonation. Rich sloping seepage fens receive calcium-rich groundwater or runoff and contain many rare and uncommon plants such as Kalm's lobelia, several species of sedges, orchids, and non-peat mosses. Patterned fens have long, linear hummocks called strings that form an alternating pattern with intervening hollows called flarks. Strings and flarks are arranged perpendicular to the direction of groundwater flow. High-elevation sloping fens of the White Mountains are unusual in that the dominant plant is Pickering's bluejoint, a grass restricted to northeastern North America.

Dominant landscape setting and vegetation structure differentiate medium and rich fen communities. Plants of relatively low stature (3 feet or less in height) dominate level sedge, shrub, and shrub–sedge fens and sloping sedge, graminoid, and shrub fens. Tall shrubs (more than 5 feet in height) and scattered saplings or trees dominate level to sloping wooded fens, and herbaceous marsh vegetation dominates marshy communities of peatland margins. Reliable indicators of the higher nutrient status of medium and

Sweet gale - meadowsweet - tussock sedge fen at Turtle Pond in Concord.

rich fens include wire sedge and other tall sedges, sweet gale, meadowsweet, marsh St. John's wort, swamp candles, cinnamon fern, winterberry, speckled alder, dogwoods, willows, and red maple. These species are sparse or absent in bogs and poor fens. In contrast to bogs and poor fens, the average pH in medium and rich fen communities is generally above 4.4.

Level Sedge, Shrub, and Shrub–Sedge Fens

This group consists of three medium fen communities that occur in level basins drained by a stream. Each community is dominated by a combination of robust sedges and medium-height shrubs (up to 3 feet). Tall shrubs and trees are absent or sparse (less than 5 percent cover). Internal streams or ponds are sometimes present within the peatland.

Sweet gale - meadowsweet - tussock sedge fen occurs statewide along the borders of streams and ponds at low to mid elevations. Sweet gale and leatherleaf dominate this medium-height, dense shrub or shrub–sedge fen community. Meadowsweet is usually present in lesser abundance, and tall shrubs are sparse or absent. Tussock sedge and other sedges and forbs are common, but collectively less abundant than shrubs. Tussock sedge, despite its name, generally does not form well-developed tussocks in this or in other peatland communities. Non-heath shrubs, sedges, and forbs help distinguish this community from the structurally similar dwarf heath shrub bogs.

Wire sedge - sweet gale fen is a widespread and common sedge or shrub–sedge fen community common on floating mats along lake and pond margins. It also occurs on soupy, loosely consolidated peat on the surface of grounded mats. Wire sedge is always abundant and mixes with sweet gale

Wire sedge - sweet gale fen at South River in Effingham.

Water willow - *Sphagnum* fen near Pawtuckaway State Park in Nottingham.

and variable amounts of other sedges such as bottle-shaped sedge and few-seeded sedge. Large cranberry is abundant below the sedges in some examples. Except for the occasional leatherleaf, other medium and tall shrubs are absent. Nutrient status is intermediate, the average pH is 4.9, and hummocks are low to moderate in height.

Water willow - *Sphagnum* fens occur along pond borders of peatlands in southern and central parts of the state. Peat mosses indicative of moderate mineral levels form a continuous carpet under the arching stems of water willow. Other plants may include silvery sedge, three-way sedge, leatherleaf, sweet gale, and swamp candles.

Sloping Graminoid and Shrub Fens

This group consists of sloping medium or rich fen communities in active groundwater seepage areas of northern New Hampshire. Sedges, grasses, or shrubs dominate. Slopes are typically gentle, but occasionally as steep as 10 degrees. A stream may drain sloping graminoid and shrub fens, but internal streams or adjacent ponds are unusual. Sloping fen communities are low in stature, dominated by herbaceous plants or medium-height shrubs (less than 3 feet tall), and moderately acidic to near neutral (average pH above 5.3). Tall shrubs and trees are absent or sparse. Half of the state's rare peatland plants are found only in two rich fen community types within this group: circumneutral - calcareous flarks and calcareous sedge - moss fens.

Montane sloping fens are a rare community type of the White Mountain region. These moderately to weakly acidic fens form on gently to moderately sloped terrain at elevations above 2,400 feet. Groundwater seepage is common, and the peat ranges from relatively shallow to moderately deep over silty material. Peat mosses are ubiquitous under a moderate cover of grasses,

Montane sloping fen on Whitewall Mountain in Bethlehem.

sedges, and forbs. Pickering's bluejoint is a dominant plant. Prickly sedge, Wiegand's sedge, and other peatland sedges are abundant, along with herbs such as false hellebore, tall meadow rue, small green woodland orchid, and tall white bog orchid.

Calcareous sedge - moss fens are sedge-dominated communities restricted to northern New Hampshire. They occur in sloped settings where calcium-rich groundwater seepage is prominent or, less often, in nearly level settings. Many rich fen forbs, shrubs, and bryophytes are present, including inland, yellow, and porcupine sedges, numerous orchids and other forbs, and a wide variety of "brown" mosses of the Amblystegiaceae family (see sidebar). Peat mosses are sparse, but when present they consist of the few species that are adapted or restricted to calcium-rich and circumneutral conditions (pH in low 7s).

Calcareous Sedge - Moss Fens

This community contains a suite of species very different from most other peatland types. Shrubs and saplings include northern white cedar (*Thuja occidentalis*) and willows (*Salix.* spp.).

Herbs include inland sedge (*Carex interior*), yellow sedge (*Carex flava*), porcupine sedge (*Carex hystericina*), chestnut sedge (*Carex castanea*), golden-fruited sedge (*Carex aurea*), slender spike-rush (*Eleocharis tenuis*), few-flowered spike-rush (*Eleocharis quinqueflora*), tawny cotton-grass (*Eriophorum virginicum*), northern cotton club rush (*Trichophorum alpinum*), red bulrush (*Scirpus microcarpus*), round-leaved sundew (*Drosera rotundifolia*), purple avens (*Geum rivale*), Robbins' ragwort (*Packera schweinitziana*), water horsetail (*Equisetum fluviatile*), variegated horsetail (*Equisetum variegatum*), marsh horsetail (*Equisetum palustre*), Kalm's lobelia (*Lobelia kalmii*), sweet coltsfoot (*Petasites frigidus* var. *palmatus*), northern green orchid (*Platanthera huronensis*), tall white bog orchid (*Platanthera dilatata*), hooded ladies' tresses (*Spiranthes romanzoffiana*), and showy lady's slipper (*Cypripedium reginae*).

Mosses include *Sphagnum warnstorfii*, *Aulacomnium palustre*, *Tomenthypnum nitens*, *Mnium* spp., *Bryum pseudotriquetrum*, *Campylium stellatum*, *Climaceum dendroides*, *Fissidens adianthoides*, and other brown mosses in the Amblystegiaceae family.

Calcareous sedge - moss fen near Lime Pond in Columbia.

Circumneutral - calcareous flark and **northern white cedar circumneutral string** near Umbagog Lake in Errol.

Only one calcareous patterned fen system occurs in New Hampshire, in the vicinity of Umbagog Lake. Patterned fens are more common in northern Maine and adjacent Canada, but few are calcareous. The system consists of two alternating, linear features called strings and flarks arranged perpendicular to the direction of groundwater flow. Strings are low peat ridges and flarks are saturated hollows on either side of the strings. Characteristic vegetation in **northern white cedar circumneutral string** communities includes northern white cedar trees interspersed among a shrub layer of leatherleaf, Labrador tea, bog rosemary, and bog willow, and various herbaceous plants and mosses. **Circumneutral - calcareous flarks** are soupy, semi-aquatic peatland environments with brown algae and sparse vascular plants including buckbean, bog rosemary, small bladderwort, and pitcher plant. Several rare rushes and sedges restricted to calcium-rich conditions are present, including moor rush and livid sedge.

Winterberry - cinnamon fern wooded fen in Raymond.

Level to Sloping Wooded Fens

This group of communities consists of tall shrub and/or medium-height shrub-dominated fens in level or sloped settings. They typically occur closer to the upland border than level sedge, shrub, and shrub–sedge fens and sloping sedge, graminoid, and shrub fen communities. Of the six communities comprising this group, the first four are largely restricted to central and southern New Hampshire and the last two occur primarily in northern New Hampshire. Each of the communities has a moderate to dense cover of tall and medium-height shrubs and a scattered tree layer. Wooded tall shrub fens often occupy the upland border zone of open peatlands, or are the dominant type in peat-filled basins without ponds.

Winterberry - cinnamon fern wooded fen is a tall shrub thicket community that generally occurs in lowlands south of the White Mountains. Well-developed hummocks and hollows are present, with the hollows dominated by peat mosses and herbs such as cinnamon fern and three-seeded sedge. Winterberry, highbush blueberry, mountain holly, speckled alder, black chokeberry, black huckleberry, and male berry are all characteristic tall shrubs.

Sweet pepperbush wooded fens are less common than the somewhat similar winterberry - cinnamon fern wooded fens, and are restricted to the southeastern part of New Hampshire. Tall and medium-height shrubs dominate, and the hollows have abundant herbs and peat mosses. Common plants include sweet pepperbush, highbush blueberry, winterberry, and coastal species such as Virginia chain fern, swamp azalea, separated sedge, and dangleberry.

Medium-height shrubs such as sweet gale and meadowsweet dominate

Sweet pepperbush wooded fen at Cedar Swamp Pond in Kingston.

Cottongrass in a **highbush blueberry - sweet gale - meadowsweet shrub thicket** at Wilkinson Brook in Effingham.

highbush blueberry - sweet gale - meadowsweet shrub thickets. Leatherleaf is always present. Taller shrubs of this community include highbush blueberry, male berry, winterberry, chokeberry, and speckled alder. Red maple trees are common. The community occurs primarily in central and southern parts of New Hampshire, but also occasionally farther north.

Medium-height and tall shrubs and robust herbs dominate **alder - lake sedge intermediate fens**. Peat mosses are abundant, and hummocks and hollows are well developed. Common shrubs include speckled alder, winterberry, male berry, witherod, highbush blueberry, leatherleaf, and sweet gale. Swamp birch, an endangered tree in New Hampshire, may be present in some examples. Robust herbs indicative of a higher nutrient status, such as lake sedge, royal fern, and skunk cabbage, distinguish this community from other wooded fens. Alder - lake sedge intermediate fens occur in central and southern parts of the state, usually in conjunction with enriched groundwater seepage.

Alder wooded fens are a broadly distributed community type that is especially abundant in the White Mountain and North Country regions. Speckled alder dominates and other shrubs are common. Scattered trees include black spruce, larch, balsam fir, and northern white cedar. The well-developed herb layer includes three-seeded sedge, silvery sedge, ferns, three-leaved false Solomon's seal, and violets.

Montane alder - heath shrub thickets resemble alder wooded fens, but feature more heath shrubs. They occur in high-elevation valleys of the White Mountains, and in lowland areas of the North Country. Speckled alder, rhodora, mountain holly, witherod, Labrador tea, and blueberries are common beneath a scattered overstory of black spruce or larch. Peat mosses are abundant, and hummock-and-hollow microtopography is moderately developed.

(Left) **Alder lake sedge intermediate fen** south of Ossipee Lake.

(Right) **Alder wooded fen** near Cascade Brook in Pittsburg.

Marshy Communities of Peatland Margins

Herbaceous emergent marsh species characterize this group of peatland communities. All three of the communities occur on peat or muck substrates, but since marsh herbs are more abundant than typical fen plants, the communities are transitional to marshes.

Floating marshy peat mats consist of thin, flat, loosely consolidated peat found along calm margins of lakes, ponds, and slow-moving streams. Species composition is variable, but often includes pond and water lilies, spike-rushes, beak rushes, sundews, bladderworts, cotton grass, St. John's

Floating marshy peat mat at Ponemah Bog in Amherst.

Marshy moat at Dead Pond in Pawtuckaway State Park.

wort, northern blue flag, and other forbs and graminoids. Stunted shrubs are either sparse or completely absent.

Marshy moats often occur in the wet zone along the upland border of peatlands in southern and central parts of the state, sometimes adjacent to moss–sedge carpet or shrub fen community types. Moats are subject to seasonally fluctuating water levels and elevated mineral inputs from upland runoff. The community is characterized by an abundance of emergent marsh grasses, bulrushes, rushes, bur reeds, and floating-leaved aquatics. These plants mix with species found in moss or moss–sedge carpet communities, such as silvery and three-way sedges.

Sedge meadow marsh is a transitional community between fen and marsh. It contains an abundance of peat mosses, sedges, and some shrubs characteristic of fens or marshes, as well as herbaceous species that are atypical of peatlands. Bottle-shaped sedge, silvery sedge, tussock sedge, three-way sedge, and bluejoint dominate. Marsh forbs are occasional to abundant. The sedge meadow marsh community is most common in wetland basins dammed and then abandoned by beavers, but it also occurs along the quiet margins of lakes, ponds, and slow-moving streams, and in the wet zone along the upland border of some peatlands.

Good Examples of Medium and Rich Fen Natural Communities

LEVEL SEDGE, SHRUB, AND SHRUB–SEDGE FENS

Sweet gale - meadowsweet - tussock sedge fen: Betty Meadows (Northwood), Wilkinson Brook (Effingham), northwest of Umbagog Lake (Errol), Little and Big Church ponds (Livermore/Albany), Berry Pond (Moultonborough/Sandwich), and Bradford Bog (Bradford).

Wire sedge - sweet gale fen: South of Ossipee Lake (Ossipee), Bearcamp River delta (Ossipee), Berry Pond (Moultonborough/Sandwich), Powwow River (Kingston), and World End Pond (Salem).

Water willow - *Sphagnum* fen: Lynxfield Pond (Chichester), Cedar Swamp Pond (Kingston), Hubbard Pond (Rindge), and Binney Pond (New Ipswich).

SLOPING GRAMINOID AND SHRUB FENS

Montane sloping fen: Shoal and Ethan Pond vicinity (Lincoln), Whitewall Mountain (Bethlehem), and North Bald Cap Mountain (Success).

Calcareous sedge - moss fen: All occur on private property in the North Country.

Circumneutral - calcareous flark: Umbagog Lake vicinity (Errol).

Northern white cedar circumneutral string: South Bay Bog (Pittsburg) and Umbagog Lake vicinity (Errol).

LEVEL TO SLOPING WOODED FENS

Winterberry - cinnamon fern wooded fen: East of Pine River (Ossipee), Mud Pond (Hillsborough), and Bradford Bog (Bradford).

Sweet pepperbush wooded fen: Spruce Swamp (Fremont).

Highbush blueberry - sweet gale - meadowsweet shrub thicket: Wilkinson Brook (Effingham), Berry Pond (Moultonborough/Sandwich), Powwow River (Kingston), and Betty Meadows (Northwood).

Alder - lake sedge intermediate fen: South of Ossipee Lake (Ossipee).

Alder wooded fen: South Bay Bog (Clarksville) and Sweat Meadow (Errol).

Montane alder - heath shrub thicket: upper East Branch of the Pemigewasset River watershed near Shoal Pond and Ethan Pond in the White Mountains, at elevations above 2,400 feet.

MARSHY COMMUNITIES OF PEATLAND MARGINS

Floating marshy peat mat: Binney Pond (New Ipswich), World End Pond (Salem), Ponemah Bog (Amherst), and Pickerel Cove (Stoddard).

Marshy moat: Trask Swamp (Alton), Town Hall Bog (Lee), Dead Pond (Nottingham), and Nye Meadow (Stoddard).

Sedge meadow marsh: Greenough Pond (Salisbury), Broad Brook (Chesterfield), Deering Wildlife Sanctuary (Deering), Orenda-Stickey Wicket Wildlife Sanctuary (Marlow), Norton Pool Preserve (Pittsburg), and Elbow Pond (Woodstock).

CHARACTERISTIC SPECIES OF SELECTED MEDIUM FEN NATURAL COMMUNITIES

A = Sweet gale - meadowsweet - tussock sedge fen
B = Wire sedge - sweet gale fen
C = Water willow - *Sphagnum* fen
D = Winterberry - cinnamon fern wooded fen
E = Alder wooded fen

COMMON NAME	SCIENTIFIC NAME	A	B	C	D	E
SHRUBS & SAPLINGS						
Large cranberry	Vaccinium macrocarpon		o			
Sweet gale	Myrica gale	•	•	o		
Leatherleaf	Chamaedaphne calyculata	•	o	o	o	
Meadowsweet	Spiraea alba	•	o			o
Water willow	Decodon verticillatus			o	o	
Red maple	Acer rubrum				o	
Winterberry	Ilex verticillata				•	
Sheep laurel	Kalmia angustifolia				o	
Other tall shrubs	Vaccinium, Lyonia, and Photinia spp.			o	•	o
Black spruce	Picea mariana				o	o
Speckled alder	Alnus incana				•	•
Balsam fir	Abies balsamea					o
HERBS						
Bluejoint	Calamagrostis canadensis	o	o			
Wire sedge	Carex lasiocarpa	o	•			
Bottle-shaped sedge	Carex utriculata	o	o			
Few-seeded sedge	Carex oligosperma		o			
Common arrowhead	Sagittaria latifolia		o			
Common cattail	Typha latifolia	o	o		o	
Tussock sedge	Carex stricta	•			o	
Swamp candles	Lysimachia terrestris	o	o	o	o	
Marsh St. John's-wort	Triadenum virginicum	o	o	o	o	
Three-way sedge	Dulichium arundinaceum			o		
Silvery sedge	Carex canescens	o		o	o	o
Cinnamon fern	Osmunda cinnamomea				•	
Crested wood fern	Dryopteris cristata					o
Three-seeded sedge	Carex trisperma					o
NON-VASCULAR						
Peat moss	Sphagnum henryense	•				
Peat moss	Sphagnum lescurii		•			
Peat moss	Sphagnum cuspidatum	o			o	
Peat moss	Sphagnum fimbriatum	•		•	o	
Peat moss	Sphagnum recurvum			•		
Peat moss	Sphagnum flexuosum			•		
Peat moss	Sphagnum fallax				•	
Other peat mosses	Sphagnum spp.					•

• = abundant to dominant   o = occasional or locally abundant   * = state threatened or endangered

MEDIUM FEN
NATURAL COMMUNITIES

POND BORDER

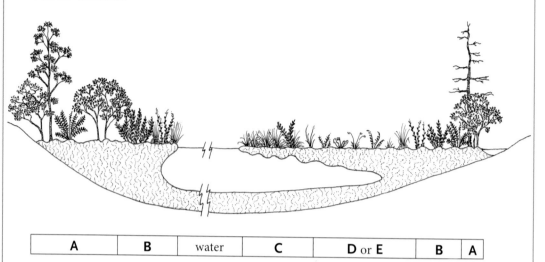

| A | B | water | C | D or E | B | A |

STREAMSIDE

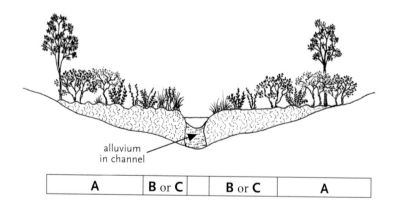

alluvium
in channel

| A | B or C | B or C | A |

A. **Winterberry - cinnamon fern wooded fen** – Border areas adjacent to swamp or upland

B. **Sweet gale - meadowsweet - tussock sedge fen** – Seasonally flooded fens adjacent to ponds or streams

C. **Wire sedge - sweet gale fen** – Seasonally flooded to saturated fen mats bordering ponds or streams

D. **Large cranberry - short sedge fen** – Saturated moss carpets with short sedges, and dwarf shrubs

E. **Bog rosemary - sedge fen** – Interior portions of the peat mat with fewer nutrients

Peatland Wildlife

The structural and compositional complexity of New Hampshire's peatlands provides habitat for a diverse suite of wildlife species. Peatland invertebrates include generalists common in all wetland habitats, such as whirligig beetles, water striders, and water boatmen, as well as many species that have specialized relationships with particular host plants found only in peatlands.

Peatland specialists include bog elfin and bog copper butterflies, whose larvae feed on black spruce and cranberry, respectively. Several spiders, assassin bugs, and ambush bugs that associate with peatland orchids eat pollinators such as bees and flies. They camouflage themselves by matching the form or color of the flower, or lie in wait inside the flower. The acidic conditions of bogs and poor fens are lethal to certain invertebrates, such as earthworms and mayflies, and the low levels of calcium prevent shell development in snails and other mollusks.

Pitcher plant is a carnivorous species that catches and dissolves insects to augment the low levels of available minerals in the nutrient-poor soils. Insects are attracted to the opening of the pitcher by visual lures and nectar. However, the pitcher plant does not trap and dissolve all invertebrates. The pitcher plant mosquito completes most of its life cycle within pitcher cavities. Mosquito eggs, larvae, and pupa successfully develop in the cavity. Several moths feed on the roots and leaves of the plant. A variety of flies, aphids, mites, and ants also feed on, or complete part of their life cycle within pitcher plants.

Ringed boghaunter, New Hampshire's only state-endangered dragonfly, breeds in medium fens of coastal and southern parts of the state. They live in fens as larvae for two to three years before emerging in mid spring as adults to hunt in nearby woods. Their early flight period precludes them from colder peatlands farther north. Peatlands are also home to common dragonflies such as blue dashers, dot-tailed whitefaces, and spotted skimmers.

The acidic water and cold, saturated conditions associated with most peatlands make them unsuitable for many amphibians and reptiles, but a few notable species prefer peatlands. The four-toed salamander will crawl to the edge of a bog and lay its eggs where *Sphagnum* moss overhangs the water. When the larvae hatch, they slide into the water. Spotted turtles utilize peatland pond shores where they frequently can be seen basking in the sun.

The relatively cool temperatures surrounding bogs and fens in northern New Hampshire provide breeding habitat for boreal bird species including olive-sided, alder, and yellow-bellied flycatchers, boreal chickadees, rusty blackbirds, spruce grouse, Lincoln's sparrow, and palm warblers. Each of these species has different habitat requirements within a peatland, and some

New Hampshire's smallest dragonfly, the elfin skimmer, is a peatland specialist.

use nearby upland forest habitat. Rusty blackbirds nest in the spruce - fir forests adjacent to peatlands. Spruce grouse also use the forest edges; peatlands provide summer fruit while shaded forest understories offer cover and safety. Palm warblers often nest on peat moss hummocks beneath conifers. At night, barred owls hunt for small peatland mammals such as water shrews. Peatlands in southern New Hampshire lack most of these species, but do provide habitat for other birds species that utilize bogs and fens.

Moose, white-tailed deer, black bear, and beaver are the most common large mammals found in peatlands. Moose browse on aquatic plants in lakes, ponds, or streams adjacent to peatlands. Bear seek blueberries and cranberries, fruits that are often abundant in these wetlands. Small mammals, including a variety of mice, voles, shrews, bats, and lemmings are frequent inhabitants of peatlands. In northern New Hampshire, northern bog lemmings feed on sedges and live in tunnels beneath the surface of drier peat mats. Their rarity, elusive nature, and remote North Country location make them one of the least understood and studied mammals in the state.

5 Swamps

Swamps are wetlands where trees and shrubs form a dense to broken canopy over herbs, mosses, and seasonal pools of water. They occur in wet depressions, along drainages, and in fringe areas transitional to marshes and peatlands. Swamps are distinguishable from upland forests by a combination of wet soils, a prominent tall shrub layer, and well-developed hummock-and-hollow microtopography. Swamps are drier than marshes and peatlands, but wetter than floodplain forests. They provide critical habitat for wildlife and plants, contribute to the abundance and diversity of the state's natural communities, and perform important ecosystem services such as water-quality improvement and flood storage.

New Hampshire's swamps are structurally and compositionally diverse, often with a well-developed understory beneath a tree canopy of 25 percent cover or more. Tall shrubs, dwarf shrubs, ferns, forbs, sedges, and bryophytes are the main life-forms in the understory. A varied microtopography in many swamps creates a range of moisture conditions that support a mix of upland and wetland species. The number of plant species in New Hampshire swamps is unknown, but more than 300 vascular plants occur in Atlantic and northern white cedar swamps alone. Twenty-six rare plant species inhabit the state's nutrient-enriched swamps, and another eighteen inhabit nutrient-poor swamps.

Swamp plants have adapted to poorly oxygenated soil conditions caused by high water levels. Shallow roots provide access to better-oxygenated surface soil. Many swamp shrubs, and some trees, absorb oxygen through large horizontal pores on their stems called lenticels. Lenticels are the dark spots on the stems of birches, winterberry, and speckled alder. Some shrubs have roots that sprout after flooding or extend vertically in response to a high

PRECEDING PAGE: A northern white cedar swamp in Northumberland.

Canopy of an Atlantic white cedar swamp in Newton.

Some hummocks and hollows in swamps originate from the windthrow of shallow-rooted trees.

water table. Seedlings develop on exposed portions of parent roots and germinate underwater. Some plants also avoid inundation by growing above the water surface on mounded hummocks. Swamp shrubs and herbs have many of the same adaptations to inundation and low oxygen conditions found in peatland and marsh plants. One important difference is that swamp plants must tolerate the reduced light conditions beneath the tree canopy.

A nurse log provides substrate for bryophytes, herbs, shrubs, and young trees.

Tree roots are often shallow in swamps since they do not need to be deep to access water, and because there is more access to oxygen close to the surface. The shallow roots make the trees vulnerable to windthrow, although some hardwoods, such as swamp white oak, resist windthrow with swollen or buttressed trunks that increase stability. When swamp trees fall, their root systems are ripped from the soil creating small depressions that hold water for extended periods. The decaying root masses and corresponding holes form hummock-and-hollow microtopography. Small-scale ground surface relief is especially common in the wettest swamps. Fallen tree trunks serve as nurse logs for germinating seeds, and provide substrate for mosses and herbs.

Many swamps are the final stage of hydrarch succession. In hydrarch succession, organic matter and sediments accumulate in lakes, ponds, and periodically flooded drainageways, raising the surface of the ground with respect to the water table, and shifting vegetation from submersed aquatic to emergent to shrub to swamp. The successional endpoint for stagnant, saturated peatlands is nutrient-poor peat swamps, whereas the successional endpoint in stream drainages is nutrient-enriched, mineral soil swamps. Conversely, beaver and human impoundments raise water levels, reversing the successional sequence, and converting swamps to marsh or fen. Snags in marshes and peatlands are evidence of former swamps or forests that were inundated by rising water. Dam abandonment or impoundment removal lowers the water level, re-initiating hydrarch succession.

The canopy composition of New Hampshire's swamps varies with latitude. Temperate broadleaf deciduous trees dominate swamps in central and southern parts of the state, whereas boreal conifers dominate northern New Hampshire swamps. Temperate hardwoods produce abundant carbohydrates during the growing season, storing the majority in their roots. The stored carbohydrates support root activity and help the tree avoid root damage during extended periods of inundation by oxygen-poor water. This strategy is ineffective in colder climates, where the growing season is shorter, water remains frozen for extended periods, and nutrients are less available. In contrast to temperate hardwoods, boreal conifers produce fewer carbohydrates, store proportionally less in their roots, and cannot support winter root-growth without suffering lethal root damage. Boreal conifers cope better with low nutrient availability, because evergreen leaves conserve nutrients and allow the tree to resume growth rapidly in the spring.

Swamp communities are separated into two broad groups based on nutrient levels. Poor swamps are very acidic and relatively nutrient-poor. They occur in isolated, poorly drained basins with organic soils, including those around the margins of open peatlands. Rich swamps are moderately acidic to weakly alkaline, better drained, and moderately to strongly nutrient-enriched. They usually have mineral soils and occur around the margins of marshes or in drained basins embedded within upland forests.

Climate and landscape setting are the primary factors affecting nutrient levels in swamps. In the colder climate of northern New Hampshire, the combination of low soil temperature, short growing season, and saturated soil reduces microbial decomposition of organic matter, and thus nutrient availability. The opposite tends to be true in southern New Hampshire.

Landscape setting modifies climate at the local scale by affecting the source and magnitude of external nutrient inputs and the amplitude of water-level fluctuations. Poor swamps occur in relatively isolated settings where nutrient inputs and water-level fluctuations are low. The saturated conditions associated with minimal water-level fluctuations in these settings impedes aeration and leads to the development of nutrient-poor organic soils. Rich swamps occur in less isolated settings affected by greater amounts of surface runoff and groundwater. The water table drops well below the surface for a portion of the growing season, producing only seasonally saturated or seasonally flooded conditions that enhance decomposition during periods of low water. As a result, most rich swamps have mineral soils with a shallow organic muck layer.

POOR & RICH SWAMP

POOR SWAMP

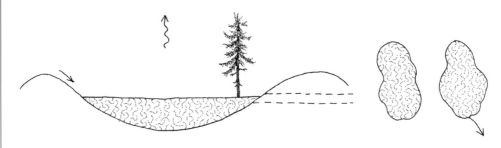

RICH SWAMP

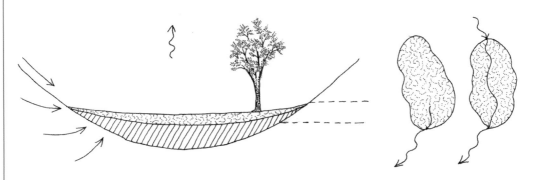

KEY:

– – – – Typical annual water level variation

——→ Water and nutrient inputs

↑ Evapotranspiration

↓ Drainage

[dotted pattern] Peat or muck soil

[hatched pattern] Mineral soil

Idealized cross-section and plane views of poor and rich swamps. **Poor swamps** *occur in isolated depressions, have relatively low inputs of water and nutrients from the surrounding landscape, and have organic peat or muck soils.* **Rich swamps** *are better drained, have greater water and nutrient inputs, broader water level fluctuations, and mineral soil with or without a shallow organic layer at the surface. NOTE: Though only one representative tree is shown for each basin, these swamps support trees all the way across.*

Poor swamps form in hydrologically isolated basins with limited nutrient inputs.

Poor Swamps

New Hampshire's poor swamps often have a forbidding, almost prehistoric appearance. Flooded hollows lie beneath a canopy of moss and lichen-covered trees with leaning trunks, gnarled roots, and scraggly upper crowns. Dense tangles of heaths and ferns spread over an undulating surface of saturated peat moss.

Poor swamp communities occur in perched depressions, stagnant headwater basins, and around the margins of open peatlands. They have relatively small watersheds and little water inflow or outflow; consequently, they receive few nutrients from external sources. Soil conditions are acidic, nutrient poor, and saturated to seasonally flooded. Organic soils, those with a surface layer of peat or muck more than 16 inches thick, develop under the saturated moisture regime.

New Hampshire's poor swamps are characterized by conifers, acid-tolerant hardwoods, heath shrubs, ferns, and peat mosses. Spruces, balsam fir, Labrador tea, and other plants with boreal or north-temperate distributions dominate poor swamps in the White Mountains and northern part of the state, though boreal conifer–dominated swamps also occur farther south, primarily around open peatlands. Poor swamps at lower elevations in central and southern parts of the state generally contain plants with temperate or coastal plain distributions. Broadly distributed temperate species include red maple, black gum, highbush blueberry, male berry, and cinnamon fern. Species with restricted temperate or coastal plain distributions are rare or uncommon in New Hampshire, and include Atlantic white cedar, pitch pine, swamp white oak, swamp azalea, sweet pepperbush, and netted chain fern.

A **black gum - red maple basin swamp** in southeastern New Hampshire.

(Left) Three-leaved false Solomon's seal is a boreal plant of poor swamps.

(Right) Netted chain fern is a rare coastal plain plant of poor swamps.

Poor swamp natural communities are structurally diverse, but species-poor. Pronounced hummock-and-hollow microtopography produces a gradient of moisture conditions from wet hollow to dry hummock. Fallen trees are common due to their shallow-rooted nature, forming mounds and nurse logs for the establishment of mosses, herbs, and woody plants. Tree, shrub, and herbaceous cover are variable; herb cover and diversity are greatest when tree cover is low. The diversity of vascular plants is typically less than in rich swamps. Peat mosses, heath shrubs, and cinnamon fern are abundant, and peatland sedges such as three-seeded sedge are common. Plants indicative of more nutrient-rich conditions, such as sensitive fern and foamflower, are sparse or absent.

Red maple, Atlantic white cedar, swamp white oak, black gum, and pitch pine dominate southern and central New Hampshire poor swamps, whereas spruce and fir dominate in the White Mountains and northward. Poor swamps in northern New Hampshire valley bottoms are frequently associated with lowland spruce - fir forests (see chapter 3). The pH of water in poor swamps ranges from the low 3s to low 5s, and averages in the mid 4s.

Temperate and Coastal Plain Poor Swamps

New Hampshire has four poor swamp types dominated by Atlantic white cedar. **Atlantic white cedar - yellow birch - pepperbush swamps** have canopies composed primarily of Atlantic white cedar and support dense thickets of sweet pepperbush and other shrubs. They occur in wet basins at low elevations within 30 miles of the coast. Tree and herb diversity is higher in

Atlantic white cedar - yellow birch - pepperbush swamp at Stratham Hill Park in Stratham.

Atlantic white cedar - giant rhododendron swamp at Manchester Cedar Swamp in Manchester.

Atlantic white cedar - leatherleaf swamp north of Country Pond in Kingston.

Inland Atlantic white cedar swamp at Loverens Mill Preserve in Antrim.

this community than in other poor Atlantic white cedar swamp communities. Frequent plants include hemlock, yellow birch, white pine, wild sarsaparilla, Canada mayflower, goldthread, and cinnamon fern. The globally rare **Atlantic white cedar - giant rhododendron swamp** occurs at fewer than ten locations in New England. The only occurrence north of Rhode Island is in Manchester. Giant rhododendron shrubs form a strikingly dense understory beneath the cedars, shading out most other plants and contributing nutrient-poor leaves to the soil. **Atlantic white cedar - leatherleaf swamps** also occur close to the coast but have an open canopy of cedar and numerous bog plants such as leatherleaf, sheep laurel, large cranberry, and pitcher plant that are uncommon in the other Atlantic white cedar swamp communities. Peat mosses are common, mossy hummocks are low, and the soils are saturated. **Inland Atlantic white cedar swamps** occur more than 30 miles from the coast, at elevations above 500 feet, and have a more northern character. They support numerous species absent in the other Atlantic white cedar communities, such as red spruce, balsam fir, bluebead lily, bunch-

Pitch pine - heath swamp at West Branch Pine Barrens Preserve in Ossipee.

Red maple - *Sphagnum* basin swamp near Pawtuckaway State Park in Nottingham.

berry, and creeping snowberry. Hummock-and-hollow microtopography is well developed and peat mosses are abundant.

Pitch pine - heath swamps are rare in New Hampshire, occurring in sand plain regions of southern and east-central parts of the state. The community is characterized by a broken canopy of pitch pine and a dense understory of tall and medium-height heath shrubs. Red maple, black spruce, gray birch, and white pine mix with pitch pine in some instances. Leatherleaf and rhodora often form a dense shrub layer, whereas herbs and peat mosses are relatively sparse. Soils are typically shallow peat over sand. Pitch pine or oak trees usually dominate the surrounding upland.

Red maple - *Sphagnum* basin swamps are common throughout central and southern New Hampshire. Red maple dominates the canopy, hemlock is often present and sometimes co-dominant, and red spruce is occasionally present in low abundance. A well-developed layer of highbush blueberry, winterberry, and other shrubs grows beneath the trees. Cinnamon fern is the most abundant herb, but marsh fern and sedges are also common. Beneath the herbs, peat mosses form a moderate to dense groundcover on hummocks and in water-filled or seasonally saturated hollows. A similar species mix occurs in **black gum - red maple basin swamps**. The significant difference between the two communities is the presence of black gum, which is dominant or co-dominant with red maple in this community (see sidebar). Black gum - red maple basin swamps are uncommon, and contain the highest concentrations of black gum trees of any habitat in New England. The community occupies perched basins on hillsides with small watersheds. Black gum is tolerant of nutrient-poor, swampy conditions, and persists by outlasting other species. The craggy, stag-headed crowns and deeply furrowed bark of old black gum are notable features of this community.

A **black gum - red maple basin swamp** in Pisgah State Park.

Black Gum

Seven hundred years ago, not long after Marco Polo returned to Venice from
China, a black gum tree sprouted from a boggy hummock in a southeast New
Hampshire swamp. Today, that individual is the oldest known hardwood tree in
North America.

Ancient black gum trees feature stout trunks, deeply furrowed bark, and
gnarled, stag-headed crowns. The angular and chaotic limbs are the product of
the tree's perpendicular branching pattern and periodic pruning by strong winds
and ice storms. Once the trees finally die, the still-standing trunks, called snags,
provide habitat for mammals and birds.

No other tree in the region has the combination of attributes so conducive
to long life that black gum does. The species grows slowly, and thus requires
few nutrients. On average, black gum adds less than a millimeter of girth each
year. It is shade tolerant, capable of growing readily after long periods of shade
suppression, and can sprout from either a clonal root system or seed. The
laterally extensive root network resulting from its clonal nature increases the
tree's resistance to windthrow. In addition, the species can tolerate flood, fire,
and drought. In short, black gum endures because it is resilient and resistant to
a broad range of environmental stresses. Importantly, black gum wood has little
timber value, an attribute that has saved the species from harvesting.

Black gum is a widespread tree of upland and lowland habitats in eastern
North America. In New Hampshire it occurs in small, nutrient-poor swamps
perched on hillside benches, along lake shores, and less often on floodplains or
in upland forests adjacent to wetlands. The species rarely attains dominance in
any community, except in swamps at the northern edge of its range here in New
Hampshire and elsewhere in southern and central New England.

A gnarled crown of
branches on an old
black gum tree at
Pisgah State Park in
Winchester.

Swamp white oak basin
swamp near Duck Pond in
Merrimack.

Swamp white oak basin swamps are restricted to depressions and flats in the coastal zone and lower Merrimack River Valley. They occur on marine sediments and other silty soils, and occasionally on sandy soils. Swamp white oak dominates the canopy, and the understory often contains cinnamon fern, highbush blueberry, and sheep laurel. Species indicative of silty soils or slightly enriched conditions include musclewood, American elm, and poison ivy. Standing water is present in the spring, but usually dries up by late summer.

Northern Poor Swamps

Two poor swamp communities occur in northern New Hampshire. **Black spruce swamp** is a conifer-dominated community of nutrient-poor basins. The community can be considered a "bog forest," and is the endpoint of the peatland successional sequence. It often surrounds open bogs or fens, and sometimes dominates entire forested basins. Black spruce is the most common canopy tree, although larch, balsam fir, and red spruce may be present in lower abundance. Canopy cover varies from open woodland with a dense heath shrub understory, to a more closed-canopy forest with a lower abundance of heaths and a greater abundance of herbs such as three-seeded sedge and cinnamon fern. Several species of peat mosses cover the ground. Soils consist of organic muck and peat material greater than 16 inches in depth.

Red spruce dominates the canopy of **red spruce swamps**, with sparser balsam fir, black spruce, and red maple. A lush understory of cinnamon fern and three-seeded sedge grows over a carpet of peat moss. Red spruce swamp communities occur on mineral soils with a shallow organic layer less than

16 inches deep, and they lack the dense heath shrub layer of more open portions of black spruce swamps. The community is common in poorly drained, perched basins in the White Mountains, and occurs less frequently in central and southwest New Hampshire.

Black spruce swamp at Lancaster Bog in Lancaster.

Red spruce swamp at Elbow Pond in Woodstock.

Good Examples of Poor Swamp Natural Communities

TEMPERATE AND COASTAL POOR SWAMPS

Atlantic white cedar - yellow birch - pepperbush swamp: Cedar Swamp Pond (Kingston), Forsaith Forest (Chester), Powwow Pond and Country Pond vicinities (Kingston), and Stratham Hill Park (Stratham).

Atlantic white cedar - giant rhododendron swamp: Manchester Cedar Swamp (Manchester).

Atlantic white cedar - leatherleaf swamp: Cedar Swamp Pond vicinity (Kingston).

Inland Atlantic white cedar swamp: Bradford Bog (Bradford), Cooper Cedar Woods (New Durham), and Loverens Mill Preserve (Antrim).

Pitch pine - heath swamp: Goodwin Town Forest (Madison), Grassy Pond (Litchfield), West Branch Pine Barrens Preserve (Ossipee), and White Lake State Park (Tamworth).

Red maple - *Sphagnum* basin swamp: Bear Brook State Park (Deerfield), LaRoche Brook vicinity (Durham), Norris Brook vicinity (Exeter), Pisgah State Park (Winchester), and Tophet Swamp (New Ipswich).

Swamp white oak basin swamp: Pickpocket Swamp vicinity (Exeter) and Tuttle Swamp (Newmarket).

Black gum - red maple basin swamp: Five Finger Point (Holderness), Fox State Forest (Hillsborough), Pawtuckaway State Park (Nottingham), Pisgah State Park (Winchester), and Stamp Act Island (Wolfeboro).

NORTHERN POOR SWAMPS

Black spruce swamp: Bradford Bog (Bradford), Cape Horn State Forest (Northumberland), Norton Pool (Pittsburg), South Bay Bog (Pittsburg), Spruce Swamp (Fremont), and Whitewall Mountain (Bethlehem).

Red spruce swamp: Annett State Forest (Rindge), Brown Ash Swamp (Thornton), Elbow Pond (Woodstock), west of Moody Ledge (Landaff), Trudeau Road vicinity (Bethlehem), and Zealand River vicinity (Bethlehem).

CHARACTERISTIC PLANTS OF SELECTED POOR SWAMP NATURAL COMMUNITIES

A = Atlantic white cedar - yellow birch - pepperbush swamp
B = Red maple - *Sphagnum* basin swamp
C = Black gum - red maple basin swamp
D = Pitch pine - heath swamp
E = Red spruce swamp
F = Black spruce swamp

COMMON NAME	SCIENTIFIC NAME	A	B	C	D	E	F
TREES							
Atlantic white cedar	*Chamaecyparis thyoides*	•					
Hemlock	*Tsuga canadensis*	o	o	o			
White pine	*Pinus strobus*	o	o	o	o		
Black gum	*Nyssa sylvatica*		o	•			
Yellow birch	*Betula alleghaniensis*	o	o	o		o	
Red maple	*Acer rubrum*	o	•	•	o	o	
Red spruce	*Picea rubens*		o	o		•	o
Pitch pine	*Pinus rigida*				•		
Black spruce	*Picea mariana*					o	•
Balsam fir	*Abies balsamea*					o	o
Larch	*Larix laricina*						o
SHRUBS							
Sweet pepperbush	*Clethra alnifolia*	•					
Northern arrowwood	*Viburnum dentatum*		o	o			
Meadowsweet	*Spiraea alba*		o	o	o		
Winterberry	*Ilex verticillata*	o	•	•	o	o	
Highbush blueberry	*Vaccinium corymbosum*	o	•	•	o		o
Mountain holly	*Nemopanthus mucronatus*	o	o	o		o	o
Sheep laurel	*Kalmia angustifolia*	o	o	o	o	o	o
Leatherleaf	*Chamaedaphne calyculata*				•		o
Rhodora	*Rhododendron canadense*				o		o
Male berry	*Lyonia ligustrina*				o		
Bunchberry	*Cornus canadensis*					o	o
Black huckleberry	*Gaylussacia baccata*				o		o
Creeping snowberry	*Gaultheria hispidula*					o	o
HERBS							
Marsh fern	*Thelypteris palustris*		o	o			
Silvery sedge	*Carex canescens*		o	o	o		
Royal fern	*Osmunda regalis*	o			o		
Cinnamon fern	*Osmunda cinnamomea*	o	•	•		•	•
Three-seeded sedge	*Carex trisperma*	o	o	o		•	•
Goldthread	*Coptis trifolia*	o	o	o		o	o
NON-VASCULAR							
Peat mosses	*Sphagnum* spp.	•	•	•	o	•	•

• = abundant to dominant   o = occasional or locally abundant   * = state threatened or endangered

Marsh marigold in bloom in early spring in a **red maple - black ash swamp**
at Shaker Village in Canterbury.

Rich Swamps

Rich swamps are diverse and productive wetlands influenced by nutrient inputs from the surrounding landscape. Red maple, black ash, American elm, and northern white cedar characterize rich swamp tree canopies. The shrub and herb layers are particularly diverse and lush. Species indicative of enrichment include speckled alder, sensitive fern, foamflower, and various grasses, orchids, and violets. Nutrient levels and pH vary among rich swamp communities. Strongly enriched, basic or weakly acidic swamps are rare in New Hampshire, whereas moderately enriched, more acidic swamps are relatively common. The richest swamps are concentrated in the Connecticut River watershed and northern parts of the state where calcium-rich bedrock is more abundant.

Rich swamps most commonly occupy depressions along headwater drainages and gentle slopes adjacent to streams, marshes, or fens. They receive external nutrient inputs from surface runoff, overbank flow, and groundwater seepage, with the level of input dependent on the mineral content of the bedrock and soil. Water inputs to rich swamps are seasonally variable, producing greater water-level fluctuations than are found in poor swamps. Periods of water drawdown aerate the soil and promote microbial decomposition of organic matter, which increases the availability of nutrients to plants. As a result, most rich swamps have mineral soils, although some have deeper muck soils.

Water source affects rich swamp hydrology patterns and plant species composition. Rich swamps adjacent to streams experience seasonal flooding from overbank flow. They often develop from former marshes and support marsh plants such as tussock sedge, bluejoint, false hellebore, and mead-

Rich swamps such as this one in the North Country are lush wetlands with high species diversity.

(Left) Small yellow lady's slipper is a very rare plant of rich swamps in northern New Hampshire.

(Right) Small purple-fringed orchid occurs in rich swamps throughout the state.

owsweet. Surface or near-surface runoff strongly influences rich swamps in drained flats, basins, or headwater areas. These swamps may be seasonally flooded or have a seasonally high water table. Plant species indicative of mineral enrichment are present, and seepage indicators and marsh plants are absent or low in abundance. Forests with seasonally high water tables occur in transition areas between uplands and swamps. They exhibit soil characteristics and plant species compositions intermediate between the two, but vegetatively they are more swamp than upland forest.

Seepage wetlands comprise a small percentage of New Hampshire's wetlands, but contain a high proportion of the state's swamp plant species. Swamps with strong groundwater seepage occupy depressions and gentle slopes, and are saturated, seasonally saturated, or flooded for short periods. Soils are well aerated. These communities are more mineral-rich than rich swamps dominated by overbank flow or surface runoff. Black ash, golden saxifrage, American pennywort, skunk cabbage, orchids, and spicebush are indicators of rich swamps strongly influenced by groundwater seepage.

Size, slope, and distribution distinguish three broad types of forested seepage wetlands: seepage swamps, seepage forests, and forest seeps. Seepage swamps form in flat to gently sloped basins ranging from less than an acre to more than 50 acres in size. Seepage forests are distinctly sloped wetlands no more than 5 acres in size where seepage zones and small upland patches mix. They are largely restricted to northern New Hampshire. Forest seeps are common but tiny, usually less than an acre in size. They form in stream headwater areas, along the upland margins of swamps, and on benches and sloping terrain of upland till or steep river-terrace sediments.

In general, seepage wetlands are larger and more common in northern New Hampshire, likely due to reduced evapotranspiration associated with the colder climate, and the abundance of silty, mineral soils.

Red maple dominates rich swamps south of the White Mountains, although black ash is prominent in seepage areas. Northern white cedar, larch, and black ash dominate swamps and seepage forests in northern New Hampshire. Rich swamps are generally sparse in the White Mountains. Herbs and shrubs dominate forest seeps, which occur throughout the state but are most abundant in northern New Hampshire. Forests with a seasonally high water table occur from the White Mountains southward. Forests in northern New Hampshire with a seasonally high water table are seepage forests or relatively wet examples of lowland spruce - fir forest (see chapter 3). The pH levels in rich swamps are mostly in the low 5s or higher.

Red maple - black ash swamp along Pressey Brook in Lyme.

Temperate Swamps

Red maple - black ash swamps occur in gently sloping, spring-fed headwater basins and areas adjacent to the upland margins of larger wetland complexes in central and southern New Hampshire. The soils are mineral or muck, red maple dominates the tree canopy, and black ash is present in low abundance. A diverse assemblage of shrubs, herbs, and bryophytes indicative of nutrient-rich seepage conditions distinguish this community from most other types of rich swamps. Characteristic species include northern spicebush, swamp saxifrage, golden saxifrage, marsh marigold, Pennsylvania bitter cress, spotted touch-me-not, small enchanter's nightshade, and several moss species. Peat mosses are infrequent, but when present include species restricted to mineral-rich conditions.

Red maple - lake sedge swamps are broadly distributed south of the White Mountains and most abundant near the seacoast. They occur around the margins of marshes on silty mineral soils with a thin muck layer. Perennial seepage or near-surface flow is prominent. Lake sedge, a tall and broad-leaved graminoid, dominates the understory, and there is a low abundance of shrubs and other herbs.

Red maple - lake sedge swamp at Wilkinson Brook in Effingham.

Red maple - sensitive fern swamps are a common type of red maple swamp that occurs on seasonally saturated or seasonally flooded mineral or muck soils. A diverse and variable assemblage of shrub and herbaceous species reflects moderately rich conditions. Characteristic plants include sensitive fern, royal fern, tussock sedge, northern blue flag, tall meadow rue, and violets. In contrast to red maple - black ash swamps and red maple - lake sedge swamps, surface or near-surface runoff is the predominant source of external nutrients. Plant species indicative of mineral-rich groundwater seepage are absent or sparse. Red maple - sensitive fern swamps vary in size,

Red maple - sensitive fern swamp in Canterbury.

Seasonally flooded red maple swamp near Crane Neck Pond in Canterbury.

Seasonally flooded Atlantic white cedar swamp at Locke Pond in Rye.

Northern white cedar - balsam fir swamp at Hurlbert Swamp in Stewartstown.

occupy depressions and drainages, and are often part of a mosaic of swamp communities.

Seasonally flooded red maple swamps, a common community type, occur adjacent to meadow marshes and shrub thickets along low-gradient streams. They are often successional from meadow marshes following beaver dam abandonment, and contain an abundance of clonal graminoids and other meadow marsh plants such as tussock sedge, bluejoint, and meadowsweet.

Seasonally flooded Atlantic white cedar swamps are uncommon in New Hampshire, occurring primarily along dammed streams in the Seacoast region and south-central parts of the state. Unlike other Atlantic white cedar swamps, this community has a relatively open cedar canopy and numerous herbaceous species typical of marshy habitats. Characteristic marsh plants that may be present include meadowsweet, bluejoint, tussock sedge, northern blue flag, and marsh St. John's wort.

Northern Swamps and Seepage Forests

Several rich swamp types dominated by hardwoods or mixed conifers, and also a number of northern white cedar swamp communities, occur in the northern part of the state. Northern white cedar is primarily a boreal tree species. It occurs frequently in northern New Hampshire, and only sparingly south of the White Mountain region. Northern white cedar is often associated with other conifers, and to a lesser extent northern hardwoods, though it can also occur in nearly pure, closed-canopy stands.

Northern white cedar - balsam fir swamp is a cedar-dominated community that develops on weakly acidic to circumneutral organic soils. In mature

Northern hardwood - black ash - conifer swamp at Coleman State Park in Stewartstown.

examples, the typical structure consists of overstory canopies reaching 40 to 60 feet high, with frequent leaning trees and blowdowns, well-developed hummock-and-hollow microtopography, sub-canopy tree and shrub layers, dense carpets of diverse mosses and liverworts, and sparse to moderate herb cover. Rare plant species, occasionally found in fen-like openings within the swamps, include several kinds of lady's slippers, sweet coltsfoot, chestnut sedge, fairy slipper, and Loesel's twayblade. **Northern white cedar seepage forests** are a similar community, but occur on sloping mineral soils between uplands and swamps on organic soil. Plants of both rich woods and seepage wetlands are present, such as foamflower, zigzag goldenrod, and Jack-in-the-pulpit.

As the name implies, **acidic northern white cedar swamps** develop on sites with more acidic conditions (water pH in the low 5s or less) than other cedar swamps, but are more enriched than poor swamps. They are associated with large, very poorly drained peatland basins around fens, and lack many of the herbaceous mineral-rich indicator species found in other cedar swamps. The herb layer is less dense and diverse, but the shrub layer is well developed. Species that help distinguish this from other northern white cedar swamps include black spruce, sheep laurel, mountain holly, creeping snowberry, and abundant peat mosses. The **northern white cedar - hemlock swamp** is a rare community found only in the Saco River watershed south of the White Mountains. It occurs in small tributary headwater basins where groundwater seepage influences the local hydrology. Cedar is dominant, and hemlock and red maple are abundant. Species found in this community that are absent from other cedar swamps include partridgeberry, white ash, trailing arbutus, and witch hazel.

Northern hardwood - black ash - conifer swamps are seepage swamps

Acidic northern white cedar swamp at Umbagog State Park in Errol.

Larch - mixed conifer swamp at Cape Horn State Forest in Northumberland.

occurring on mineral or shallow peat soils. The most common tree species in the diverse overstory are black ash, yellow birch, balsam fir, and red spruce. The herb layer is lush and indicative of mineral-rich conditions. Common herbaceous species include purple avens, Robbins' ragwort, water pennywort, and small enchanter's nightshade. Several rare plant species sometimes occur in this community, including green adder's mouth and heart-leaved twayblade.

Larch - mixed conifer swamps are saturated conifer swamps intermediate in enrichment between northern white cedar - balsam fir swamps and nutrient-poor black spruce swamps. Larch and balsam fir dominate the canopy, with variable amounts of red, black, and white spruces, northern white cedar, and hardwoods such as red maple and black ash. Herbs include cinnamon fern, northern wood sorrel, dwarf raspberry, and three-seeded sedge.

Northern hardwood seepage forests are common in the northern part of the state. These semi-rich wetlands occur on lower mountain slopes that have frequent seep openings and seepage runs. Common trees include sugar maple, yellow birch, and occasionally black ash. The herb layer is lush and diverse, and often contains northeastern mannagrass, spotted touch-me-not, purple-stemmed aster, and dwarf raspberry. In some examples, the seepage zones occupy extensive areas, with lush glade openings that can reach a half-acre in size.

Northern hardwood seepage forest in Gray Wildlife Management Area in Pittsburg.

Forest Seeps

Forest seeps are very small wetlands, typically less than 1 acre in size, that occur around groundwater discharge areas in upland forests. Species com-

(Left) A small **acidic *Sphagnum* forest seep** on North Twin Mountain in Bethlehem.

(Right) **Subacid forest seep** at Northwood Meadows State Park in Northwood.

position varies among sites, but collectively seeps are diverse and support a distinctive flora and fauna. Despite their small size, seeps add a distinct biological component to the matrix of upland forests. Many herbs, sedges, and mosses are restricted to forest seeps or other seepage wetlands.

The primary feature used to classify seeps is acidity level, which is indicated by corresponding plant species assemblages. **Acidic *Sphagnum* forest seeps** are the most acidic seep type, occurring in high- and low-elevation coniferous forests, on wet outcrops, and along the upper reaches of cool, entrenched streams. Peat mosses dominate. Species composition is variable and may include false hellebore, whorled aster, cinnamon fern, and several sedge species. **Subacid forest seeps** exhibit weakly acidic conditions and are highly variable in species composition. Common plants of these seeps include foamflower, rough sedge, small enchanter's nightshade, and golden saxifrage. Subacid seeps occur throughout the state, with pH ranges from the mid 4s to 6. **Circumneutral hardwood forest seeps** have neutral or weakly acidic soil water, with pH generally higher than 6, and are less common than subacid seeps. They contain rich mesic hardwood species such as wood nettle, ostrich fern, zigzag goldenrod, and plantain-leaved sedge, and occasionally support populations of rare plants such as yellow lady's slipper and Loesel's twayblade.

Circumneutral hardwood forest seep at St. Gaudens National Historic Site in Cornish.

Hemlock - cinnamon fern forest at Northwood Meadows State Park.

Red maple - red oak - cinnamon fern forest at Wagon Hill Farm in Durham.

Forests with Seasonally High Water Tables

These wet forests occur in transitional areas between swamps and upland habitats and may or may not meet the legal criteria for wetlands. They are seasonally saturated, or have a seasonally high water table, with soils ranging from poorly drained to moderately well drained. In northern New Hampshire, forests with a seasonally high water table are usually wet examples of lowland spruce - fir forest (see chapter 3).

Hemlock - cinnamon fern forests occur along gently sloped drainages and in poorly drained areas of upland forests. Hemlock and red maple dominate the tree canopy, although other tree species may be present. The understory of cinnamon fern, marsh fern, and various mosses, including peat mosses, indicates mesic to wet conditions. They mostly occur south of the White Mountains.

Red maple - red oak - cinnamon fern forests are dominated by hardwood trees, particularly red maple, red oak, and birches, with sparse pine and hemlock. Cinnamon fern and tall wetland shrubs such as highbush blueberry are present in moderate abundance. Other wetland plants are sparse. The community is similar to hemlock - cinnamon fern forest, but hardwoods dominate and the shrub and herb layers may be denser.

Red maple - elm - lady fern silt forests occur on silt or clay soils with a high water table in the Great Bay area of southeastern New Hampshire. Red maple dominates, and American elm and other trees are present in lower abundance. Musclewood can be an abundant understory species, and semi-rich site indicators such as lady fern, sensitive fern, and violets are usually present.

Red maple - elm - ladyfern silt forest at Great Bay National Wildlife Refuge in Newington.

Good Examples of Rich Swamp Natural Communities

TEMPERATE SWAMPS

Red maple - black ash swamp: College Woods (Durham), Pawtuckaway State Park (Nottingham), and Canterbury Shaker Village (Canterbury).

Red maple - lake sedge swamp: Great Meadows (Exeter), Contoocook River (Peterborough), Great Bog (Portsmouth), and Turtle Pond (Concord).

Red maple - sensitive fern swamp: Great Bog (Portsmouth) and Northwood Meadows State Park (Northwood).

Seasonally flooded red maple swamp: Northwood Meadows State Park (Northwood) and Bear Brook State Park (Deerfield).

Seasonally flooded Atlantic white cedar swamp: Fairhill Swamp (Rye).

NORTHERN SWAMPS AND SEEPAGE FORESTS

Larch - mixed conifer swamp: Cape Horn State Forest (Northumberland).

Northern white cedar - balsam fir swamp: Hurlbert Swamp Preserve (Stewartstown), Cape Horn State Forest (Northumberland), Brundage Forest (Pittsburg), and 13-Mile Woods (Cambridge).

Northern white cedar - hemlock swamp: White Horse Ledge vicinity (Albany).

Northern hardwood - black ash - conifer swamp: Coleman State Park (Stewartstown) and Umbagog State Park (Errol).

Acidic northern white cedar swamp: Whaleback Ponds (Wentworth's Location) and Umbagog National Wildlife Refuge (Errol).

Northern hardwood seepage forest: Connecticut Lakes Headwaters Natural Area (Pittsburg) and Franconia Notch (Lincoln).

Northern white cedar seepage forest: Mud Pond vicinity (Pittsburg).

FOREST SEEPS

Acidic *Sphagnum* forest seep: Trudeau Road vicinity (Bethlehem), above Crystal Cascade (Pinkham's Grant), along the Imp Trail (Bean's Purchase), and Nancy Brook Research Natural Area (Livermore).

Subacid forest seep: Allard Brook near Swift River (Albany), Edward MacDowell Reservoir (Peterborough), and Connecticut Lakes Headwaters Natural Area (Pittsburg).

Circumneutral hardwood forest seep: Jeffers Mountain and Black Mountain (Benton) and Crommet Creek vicinity (Durham).

FORESTS WITH A SEASONALLY HIGH WATER TABLE

Hemlock - cinnamon fern forest: Allard Brook vicinity along the Swift River (Albany), Bear Brook State Park (Allenstown), Northwood Meadows State Park (Northwood), and Exeter River (Exeter).

Red maple - elm - lady fern silt forest: Bellamy River Wildlife Sanctuary (Dover), College Woods (Durham), south side of Great Bay (Greenland), and Great Bay National Wildlife Refuge (Newington).

Red maple - red oak - cinnamon fern forest: Great Bay National Wildlife Refuge (Newington) and Wagon Hill Farm (Newington).

CHARACTERISTIC PLANTS OF SELECTED RICH SWAMP NATURAL COMMUNITIES

A = Red maple - black ash swamp
B = Red maple - sensitive fern swamp
C = Seasonally flooded red maple swamp
D = Northern hardwood - black ash - conifer swamp
E = Northern white cedar - balsam fir swamp
F = Larch - mixed conifer swamp

COMMON NAME	SCIENTIFIC NAME	A	B	C	D	E	F
TREES							
Red maple	*Acer rubrum*	•	•	•	o	o	o
Black ash	*Fraxinus nigra*	•			•	o	o
Yellow birch	*Betula alleghaniensis*	o			o	o	
Northern white cedar	*Thuja occidentalis*				o	•	o
Balsam fir	*Abies balsamea*				o	•	•
Red spruce	*Picea rubens*				o	o	o
Larch	*Larix laricina*				o	o	•
SHRUBS							
Poison sumac	*Toxicodendron vernix*		o				
Highbush blueberry	*Vaccinium corymbosum*	o	o	o			
Northern spicebush	*Lindera benzoin*	o	o				
Poison ivy	*Toxicodendron radicans*	o	o		o		
Northern arrowwood	*Viburnum dentatum*		o	o			
Meadowsweet	*Spiraea alba*		o	o	o		
Speckled alder	*Alnus incana*	o	o	o	o	o	o
Red osier dogwood	*Cornus sericea*	o			o	o	
Winterberry	*Ilex verticillata*		•	o	o	o	o
Witherod	*Viburnum nudum*				o	o	
Bunchberry	*Cornus canadensis*				o	o	o
Mountain holly	*Nemopanthus mucronatus*				o	o	o
HERBS							
Swamp saxifrage	*Saxifraga pensylvanica*	o					
Marsh marigold	*Caltha palustris*	o					
Brome sedge	*Carex bromoides*		o				
Sensitive fern	*Onoclea sensibilis*	o	•		o		
Spotted touch-me-not	*Impatiens capensis*	o	o		o		
Purple avens	*Geum rivale*	o			o		
Water pennywort	*Hydrocotyle americana*	o			o		
Tussock sedge	*Carex stricta*		•	•	o		
Bluejoint	*Calamagrostis canadensis*	o	•				
Cinnamon fern	*Osmunda cinnamomea*	o				o	o
Yellow lady's slipper*	*Cypripedium parviflorum**				o	o	
Twinflower	*Linnaea borealis*					o	o
Naked miterwort	*Mitella nuda*					o	
Three-seeded sedge	*Carex trisperma*					o	o

• = abundant to dominant   o = occasional or locally abundant   * = state threatened or endangered

Swamp Wildlife

In swamps, the diversity of tree species and variations in hydrology produce conditions that support a broad array of wildlife species. Bird and
small mammal species richness increases with higher structural complexity and associated food availability. Little is known about swamp invertebrates, although many insects have specific relationships with host plant
species. Reptiles and amphibians are abundant and diverse in swamps, due
to the presence of seasonal or permanent water. Amphibians travel through
swamps in search of vernal pools (see sidebar). Migrating salamanders and
frogs provide food for raccoons and other predators.

The dense shrub layer of some hardwood swamps provides high-quality
nesting habitat for birds, including black-capped chickadees, common
yellowthroats, and black-and-white warblers. Gray catbirds nest in the
densest thickets, whereas yellow warblers prefer sparser areas. Northern
waterthrush is one of the few bird species restricted to swamp habitat.
They occur in northern white cedar swamps in the north and red maple

Vernal Pools

Vernal pools are small depressions that fill and hold water for at least two
months in spring, then dry out during the summer. They have no inlets or
outlets, and occur as patches within other habitats, including upland forests,
floodplain forests, swamps, and sand plain basin marshes. Periodic natural
drawdown and hydrologic isolation prevent the establishment of fish populations. With fish predators excluded, vernal pools serve as important feeding and
breeding grounds for reptiles, amphibians, and invertebrates, several of which
are adapted to, and completely dependent on, the cyclic and ephemeral nature
of these wetland basins.

Amphibians migrate to vernal pools from their upland habitats on the first
warm rainy night of spring, when the temperature is about 40 degrees Fahrenheit (usually in early April). Characteristic amphibians include wood frogs, and
spotted, blue-spotted, and Jefferson's salamanders. These amphibians mate and
lay eggs in the pools, then return to the forest. After hatching, successful growth
and metamorphosis from larvae to adult must happen before water in the
pools disappears. Wood frogs can develop in as few as 115 days, while spotted
salamanders require over 155 days. As a result, ponds that dry by the end of
July can support wood frogs, but probably not spotted salamanders. The rare
marbled salamander, found only in the southern part of the state, lays its eggs in
the bottoms of dry vernal pools in the fall and remains with the eggs throughout

the winter until the pools fill with water in the spring. The eggs hatch, and the developing larvae eat newly hatched wood frog and spotted salamander larvae.

Other animals use vernal pools as well. Fairy shrimp lay desiccation-resistant eggs in leaf litter on the bottom of pools. In early spring, the eggs hatch and colorful orange, yellow, and turquoise crustaceans appear, swimming on their backs. Fingernail clams seal themselves tightly to survive the dry period. Many insects, including mosquitoes, have eggs or larvae that can withstand the seasonal drawdown by remaining in the moist leaf litter.

Vernal pools are unvegetated to moderately vegetated, and defined by characteristic animal species, not vegetation. Flora is often restricted to elevated mounds or areas near the pool margins. Plant species richness and cover are highly variable and depend on the extent of water fluctuations, light, geographic location, variation in seed dispersal, and other factors. More than 400 plants occur in vernal pools in northeastern North America, including common species such as highbush blueberry, buttonbush, and several graminoids.

The loss of the tree canopy around vernal pools can have a harmful impact on wildlife species. The water temperature rises with increased exposure, and some animals cannot survive in the warmer water. In addition, the amount of leaf litter in the pool decreases, eliminating a critical component of the food web. Buffer zones can help protect vernal pools from human activities in the surrounding landscape.

A vernal pool at Pawtuckaway State Park.

swamps farther south. Migratory birds, including many species of warblers, use swamps on the way to breeding grounds.

Many common upland animals use swamps during the winter for food and cover. In southern New Hampshire swamps, the seeds and berries of some shrubs and trees, such as red maple, blueberry, winterberry, and black gum, offer forage for voles, shrews, mice, and chipmunks. Deer and moose seek shelter in dense vegetation. Spotted turtles occur in Atlantic white cedar swamps, black gum swamps, and other poor swamps of southern and central parts of the state, particularly those with relatively shallow surface water. Hessel's hairstreak, a rare butterfly, spends most of its life cycle high in the canopy of Atlantic white cedar swamps.

Poor swamps generally provide low-quality browse because of the low nutrient content of conifer trees and heath shrubs. Moose, however, can extract nourishment from these plants more effectively than most herbivores, explaining in part their abundance in northern climates. Spruce grouse, olive-sided flycatcher, black-backed woodpecker, boreal chickadees, four-toed salamander, wood frog, and green frog are some of the other wildlife species associated with poor swamps in northern New Hampshire.

Few animals are rich swamp specialists. The rare fen ant, however, only occurs in rich swamps in coastal and southern New Hampshire, where it builds distinctive mounds from sedge stems and sediment. In late winter, wild turkey and other birds feed on the stalks of sensitive fern, a common plant in rich swamps and forest seeps. Seeps provide feeding and breeding grounds for amphibians and small mammals such as the southern red-backed vole. In addition, seeps often remain open throughout the winter, providing a rare source of forage in an otherwise barren forest.

6 Marshes

F reshwater marshes are herbaceous and shrubby wetlands that occur along low-gradient streams and in other lowland areas. Shallow water covers the ground surface with a frequency and duration sufficient to exclude trees or reduce tree cover to less than 25 percent. Fluctuating water levels and nutrient inputs from surrounding uplands create conditions for vigorous growth of flood-tolerant herbs. Marshes provide essential habitat for a variety of plants and animals, and valuable ecosystem services such as flood reduction, water-quality improvement, and water-supply storage.

Marshes develop in landscape settings subject to substantial seasonal variation in water level. Watershed runoff concentrates in streams, but during periods of snowmelt and high rainfall, water volume can exceed the capacity of the stream channel and wetland soil, resulting in temporary flooding. Water levels drop substantially in summer when water inputs are lower, and when plants are assimilating water and releasing it to the atmosphere. Seasonally variable water levels also occur in certain isolated, ponded depressions that are subject to large vertical groundwater fluctuations.

The combination of water-level fluctuations and seasonal inflow can produce nutrient-rich conditions. Organic matter decomposes during water-level drawdown periods, releasing nutrients unavailable under continuously saturated or inundated conditions. The result is a highly decomposed organic material called muck. Muck ranges from pure organic matter to mixes

PRECEDING PAGE: Tussock sedge and New York aster in a **tall graminoid meadow marsh** at Turtle Pond in Concord.

A marsh community system at Garland Pond Wildlife Management Area in Ossipee.

(Left) Hop sedge is frequent in nutrient-rich marshes of floodplain oxbows.

(Right) Narrow-leaved gentian occurs in a variety of marshes in New Hampshire.

of plant material and mineral sediments brought in by floodwaters. Groundwater, water flowing into marshes from surface runoff, and overbank flow from streams or lakes contain dissolved nutrients leached from soil and bedrock in the surrounding landscape. This elevated nutrient level promotes the development of marsh vegetation, primarily herbs and non-heath shrubs. In contrast, stagnant and nutrient-poor conditions in bogs and fens favor peat mosses and heath shrubs.

Three major types of marshes occur in New Hampshire, distinguished by their connection to (or isolation from) a drainage network, and the presence or absence of wave disturbance. Drainage marshes occupy the borders of streams, rivers, ponds, lakes, and other flat areas drained by streams, including oxbow basins along rivers. Periodically inundated by low-energy (slow-moving) floodwater, they experience little or no wave disturbance. Nutrient inputs are relatively high, which in conjunction with fluctuating water levels, creates productive conditions for the growth of robust herbs and shrubs. Most marshes in New Hampshire are drainage marshes.

Sand plain basin marshes occur in isolated depressions that lack inlet and outlet streams. Surface water and associated nutrient inputs are minimal, but water levels in the porous sandy soils experience significant seasonal and annual fluctuations. Low-nutrient conditions limit the growth and productivity of plants. The small amount of organic matter present decomposes rapidly during drawdown periods. Consequently, organic matter accumulation is minimal.

Sandy pond shore marshes occur along wave and ice-disturbed shores of shallow, sandy ponds and lakes, some of which have naturally fluctuating

Many marshes exhibit a zonation of communities corresponding to water depth. Aquatic bed, emergent marsh, and meadow marsh zones are seen here.

water levels. Organic matter build-up is more limited in these marshes than in drainage marshes due to low productivity, rapid decomposition, and the effects of regular wave and ice-scour disturbance.

As many as four broad zones may occur within any given marsh system, corresponding to flood regime and dominated by differing life-forms. These four zones are shrub thickets, meadow marshes, emergent marshes, and aquatic beds. Within each of the zones, differences in setting, water source, nutrients, hydrologic regime, and species composition determine the natural community. Shrub thickets are seasonally saturated to seasonally flooded and occur adjacent to wooded swamps or upland forests. Tall deciduous shrubs such as winterberry, silky dogwood, alder, and highbush blueberry dominate, and may mix with medium-height shrubs and red maple saplings. Meadow marshes (or wet meadows) are wetter than shrub thickets, but water levels remain below the surface for much of the growing season. Sedges, grasses, and forbs dominate. Emergent marshes are wetter than shrub thickets and meadow marshes. They remain inundated by shallow to deep water for most or all of the growing season. Emergent marshes contain spongy-tissue plants that emerge from the water (also called emergents), such as bur reeds, pickerel weed, and cattails. Aquatic beds are usually inundated by several feet or more of water. Floating and submersed aquatic plants, such as bladderworts, water lilies, and pondweeds dominate aquatic beds. A depth of ten feet is the outer limit for most rooted plants.

Oxygen deprivation is the primary stress that all marsh plants must endure. Flooding creates a low-oxygen or anoxic (no oxygen) environment for roots, and in deeper water, for stems and leaves. Plants must avoid or tolerate low-oxygen or anoxic conditions to prevent tissue damage or death. One plant adaptation is the concentration of roots close to the surface. Some species, such as willows, even send vertical roots to the surface to access air. Furthermore, many wetland shrubs have multiple trunks, with abundant oxygen-intake structures called lenticels.

Emergent and aquatic species experience the longest periods of inundation. Both types of plants typically have aerenchyma, a spongy tissue that helps transport oxygen to the roots via a network of gas-filled chambers. Submersed and floating-leaved aquatic plants are the most oxygen stressed, must grow the farthest to reach the surface, and have the least amount of tissue exposed to the air. Their primary strategy is to extend fast-growing, narrow, submersed stems and leaves to the surface. Once at the surface, the leaves widen. Another adaptation of aquatic plants to oxygen deprivation is their ability to generate energy through anaerobic fermentation of root-stored carbohydrates. Aquatic plants accumulate abundant fermentable carbohydrates in their roots during the summer to survive long periods of winter inundation in low-oxygen or anoxic conditions.

Short graminoid - forb emergent marsh/mudflat at Peverly Meadows in Canterbury.

Drainage Marshes

The dense growth of herbs and shrubs in drainage marshes is a common sight along the margins of New Hampshire's streams, rivers, lakes, and ponds. Drainage marshes are diverse wetlands, comprised of shrub thickets, meadow marshes, emergent marshes, and aquatic beds. Surface and groundwater nutrients facilitate the decomposition of plant remains during periods of low water, resulting in productive conditions and a seasonal flush of growth.

Drainage marshes form in ponded environments where a sequential change in wetland vegetation occurs as organic matter and sediment accumulate. The sequence, called hydrarch succession, begins with the establishment of aquatic and emergent plants in shallow areas along the water's edge. Water depth diminishes as plant remains slowly accumulate on the bottom or around the margins. Vegetation then shifts from aquatic bed and emergent marsh to meadow marsh and shrub thicket, eventually arriving at forested swamp—the end-point of hydrarch succession. The process is similar to the formation of bogs, in which thick layers of peat accumulate in ponds. Unlike bogs, however, the fluctuating water levels and seasonal inflow that characterize drainage marshes lead to rapid decomposition of herbaceous plants and shallower accumulations of highly decomposed muck soil.

Beaver and human impoundments typically reverse hydrarch succession to more aquatic conditions. Standing dead trees called snags are testament to the activity of beavers and cycles of vegetation change in drainage marshes and other wetlands. Beaver dam abandonment, and subsequent leakage or draining, re-initiates the sequence. Short-term cycles of forward and back-

Drainage marshes occur in a variety of nutrient-rich settings, such as oxbows along floodplains, like this one near the Blackwater River in Salisbury.

Northern blue flag is a common plant of many drainage marshes.

White waterlily is a common plant of aquatic bed communities.

ward successional changes in marshes are part of the long-term progression of wetland development that has occurred since the retreat of glaciers more than 10,000 years ago.

Drainage marshes support diverse, broadly distributed, temperate climate plants; plants with northern or coastal plain distributions are rare. The nutrient-enriched conditions of drainage marshes favor tall and vigorous clonal species. Clonal plants spread primarily by vegetative means, and form dense root mats. Bluejoint, tussock sedge, bulrushes, and pickerel weed are examples of plants that dominate drainage marshes. These species often create a thick litter or thatch layer that impedes the germination of plants that rely on reproduction from seed. New Hampshire's drainage marshes contain

Smooth bidens at Holts Pond in Nashua.

125 rare plant species, including 35 aquatic, 20 emergent, and 70 meadow marsh plants. The federally threatened northeastern bulrush occurs in only a handful of emergent and meadow marshes near the Connecticut River.

Shrub Thickets and Shrublands

Most of New Hampshire's shrub communities associated with marshes flood for relatively short periods compared to herbaceous marshes. All have at least 25 percent shrub cover, and less than 25 percent tree cover. Herb cover is variable. Shrub thickets have at least 60 percent shrub cover, and less than 25 percent tree cover. Four shrub communities occur on mineral soil in low-energy settings, usually with a thin muck layer at the surface.

Highbush blueberry - winterberry shrub thickets are common and widespread in central and southern parts of the state. They occur along the upland margins of streamside marshes, and surround or fill basins in sand plain and glacial till settings that lack or have restricted outlets. Highbush blueberry and winterberry form tall, dense thickets, sometimes with a low abundance of other shrub species and red maple saplings. Herbs are scarce, but include cinnamon and royal ferns. Peat mosses and other peatland plants are noticeably scarce or absent, distinguishing this shrub thicket community from similar peatland communities such as highbush blueberry - mountain holly wooded fen and winterberry - cinnamon fern wooded fen.

Alder seepage thickets occur in relatively nutrient-enriched settings adjacent to streamside marshes, or in depressions and slopes fed by ground-water seepage. Alder dominates a diverse and moderately well-developed herbaceous layer of ferns, forbs, sedges, and grasses indicative of min-

Highbush blueberry - winterberry shrub thicket at Turtle Pond in Concord.

(Above) **Alder seepage thicket** at Great Bay National Wildlife Refuge in Newington.

(Above right) **Buttonbush shrubland** in an ox bow along the Blackwater River in Salisbury.

eral enrichment or groundwater seepage. Characteristic plant species include marsh marigold, golden saxifrage, spotted touch-me-not, sensitive fern, perfect-awned sedge, mannagrasses, bluejoint, and several sedges and mosses. Mineral soils with a shallow muck layer are typical of this community.

Alder alluvial shrublands typically occur on shallow muck soils in seasonally flooded, low-energy areas adjacent to meadow marshes, and less often in higher-energy settings such as alluvial floodplains and riverbanks. In drainage marsh settings, speckled alder dominates, and a variety of herbaceous plants occurs in the understory. The community often succeeds meadow marsh communities following beaver dam abandonment and a drop in water level.

Buttonbush shrublands are most abundant in central and southern parts of the state. They form in basins, abandoned oxbow channels on floodplains, and other stagnant, semi-permanently flooded settings. The presence of standing water throughout much of the growing season limits shrub and herb diversity. Buttonbush, perhaps the most flood-tolerant deciduous shrub in New Hampshire, is the dominant plant in this community. Pioneer mosses often grow on the stems of buttonbush near the water line.

Meadow Marshes

Meadow marshes are seasonally saturated to seasonally flooded communities characterized by wetland grasses, sedges, rushes, and forbs (broadleaved herbs). Flooding may occur during the spring or high-runoff events but, in most years, the water table remains below the ground surface for much of the growing season. Some meadow marshes remain inundated by shallow water in wetter years, and are hydrologically transitional to emergent marshes. During the growing season, water levels remain substantially below the surface in mixed tall graminoid - scrub-shrub marsh, tall graminoid meadow marsh, and short graminoid - forb meadow marsh/mudflat communities, but closer to the surface in sedge meadow marsh, herbaceous seepage marsh, and lake sedge seepage marsh communities. Seventy rare plants occur in New Hampshire meadow marsh communities.

Mixed tall graminoid - scrub-shrub marsh is a common meadow marsh type, transitional in structure, hydrology, composition, and location between shrub thicket and herb-dominated meadow marsh communities. Herbaceous plants such as bluejoint, tussock sedge, bottle-shaped sedge, cinnamon and royal ferns, and Joe-Pye weeds constitute 50 percent or more of the cover. Shrubs, including highbush blueberry, winterberry, speckled alder, meadowsweet, and willows, constitute up to 50 percent of the vegetation cover. When shrub cover is greater than 25 percent, this community can, in a strict sense, be considered a type of shrubland.

Tall, clonal grasses or sedges, such as bluejoint, tussock sedge, bulrushes, and reed canary grass dominate seasonally flooded **tall graminoid meadow marshes**. These plants often form dense root mats, and may dominate individually or in combinations depending on hydrologic regime. Other species common in these marshes include mannagrasses, rice cutgrass, rushes, Joe-Pye weeds, and sedges such as lurid sedge, three-way sedge, and various spike-rushes. This community is common throughout the state.

Short graminoid - forb meadow marsh/mudflats occur along seasonally flooded stream shores or on exposed mudflats of recently abandoned beaver ponds. Short, herbaceous plants dominate, including numerous annuals and clumped or spreading perennials. For example, rice cutgrass competes successfully by spreading across the ground surface and rooting from leaf nodes on the horizontal stem. Other characteristic plants include mannagrasses,

(Left) **Mixed tall graminoid - scrub-shrub marsh** in the Marston Brook vicinity in Effingham.

(Right) **Tall graminoid meadow marsh** along the Belmont River in Belmont.

Short graminoid - forb meadow marsh/mudflat along the Merrimack River in Boscawen.

false pimpernel, goldenpert, beggar-ticks, St. John's wort, small spike-rush, and water parsnip. As inundated areas are exposed, plants emerge from buried seeds or seeds recently deposited on the surface by floods. Conditions that favor establishment of short graminoid - forb meadow marsh/mudflats are frequently temporary, and the vegetation often succeeds to tall graminoid meadow marsh communities.

Sedge meadow marshes occur in seasonally flooded or seasonally saturated areas where water-level fluctuations are relatively moderate compared to the previously described meadow marsh communities. Water levels remain close to the ground surface throughout the growing season. These sedge or sedge–grass meadows are transitional between marsh and fen, with emergent or meadow marsh plants mixing with fen sedges and peat mosses. Dominant plants include a combination of bottle-shaped sedge, silvery sedge, tussock sedge, prickly sedge, and peat mosses. Bluejoint may be common, but it is less abundant than the sedges. Common settings for sedge meadow marshes are relatively stagnant headwater basins, hydrologically isolated lobes of larger drainage marshes, and abandoned beaver impoundments. All of these are areas where lowered water levels allow for the accumulation of organic matter. Sedge meadow marshes also occur on grounded peat mats in seasonally flooded portions of medium fens.

Herbaceous seepage marshes occur in slightly sloping groundwater discharge areas on poorly drained marine sediments and other silty soils between emergent marshes and seepage swamps or shrub thickets, and along seepy headwater sections of streams. Tall forbs and ferns dominate, and species composition is variable. Dominant species include sensitive fern, marsh fern, marsh marigold, Joe-Pye weed, skunk cabbage, scouring rush, and a

A **sedge meadow marsh** dominated by bottle-shaped sedge near Greenough Pond in Salisbury.

(Left) **Herbaceous seepage marsh** at Great Bay National Wildlife Refuge in Newington.

(Right) **Lake sedge seepage marsh** at Turtle Pond in Concord.

variety of other ferns and sedges. Mosses may be abundant, although peat mosses are generally absent.

Lake sedge seepage marshes are a narrowly defined, but relatively common type of seepage marsh occurring as small patches or bands between emergent marsh and swamp or upland. Lake sedge, a tall plant that spreads by rhizomes, dominates these slightly sloping marsh areas that are saturated by perennial near-surface flow or have reliable groundwater seepage. Settings with this hydrology also include the inflow and outflow sections of marshes. Soils are silty, with a shallow muck layer under a thick thatch of dead sedge leaves. Plant species indicative of the mineral-enriched conditions found in herbaceous seepage marshes may also be present, but they are much less abundant than lake sedge.

Emergent Marshes

Emergent marshes are saturated to semi-permanently flooded wetlands with seasonally variable water levels. They occur throughout the state and are characterized by herbaceous plants with spongy tissue (aerenchyma) emerging above the surface of the water. Twenty rare plant species occur in New Hampshire's emergent marshes, including several bur reeds, bulrushes, and spike-rushes. Narrow-leaved cattail and common reed are invasive exotics that occur in some of New Hampshire's emergent marshes. A rare native subspecies of common reed, easily confused with the invasive exotic subspecies, occurs on the upper reaches of some tidal marshes.

Common plants in the typical **emergent marsh** include bur reeds, com-

(Left) An **emergent marsh** at the Thompson Sanctuary in Sandwich.

(Right) **Cattail marsh** in Deerfield.

mon arrowhead, arrow arum, pickerel weed, water bulrush, soft-stemmed bulrush, three-way sedge, and common cattail. The combination of species at a given location depends on factors such as water depth and amplitude of water-level fluctuations. In shallow emergent marshes, several inches to a foot of water inundate the ground for most of the growing season. Dry periods expose shallow emergent marsh soils, but they remain saturated for the remainder of the year. In deeper emergent marshes, 2 to 3 feet of water semi-permanently inundate the ground. Only severe droughts expose the soils. In addition to emergent plant species, submersed and floating-leaved species are also present.

Bayonet rush emergent marshes occur in shallow ponds, lakes, and rivers with sandy bottoms or with thin organic layers over sand. They remain inundated for most or all of the growing season, although dry periods expose portions of the community. Associates of the dominant bayonet rush include pickerel weed, water lobelia, bur reeds, three-square rush, goldenpert, pipewort, water shield, white waterlily, bladderworts, and pondweeds.

Cattail marshes are a common type of emergent marsh found at low elevations in New Hampshire. They occur in drained basins, around ponds, and along small streams in saturated or shallowly inundated soils. Cattails dominate the vegetation, and often exclude other species entirely.

Aquatic Bed

The **aquatic bed** community occurs in permanently inundated parts of rivers, streams, and ponds shallow enough to support rooted vegetation. Floating-leaved and submersed plants are the dominant life-forms. Water depths are usually 2 to 3 feet at annual low water, and may be as deep as 10 feet, the approximate depth limit for the growth of rooted plants. Water shield, floating heart, water flaxseed, and various species of pond lilies, water lilies, bladderworts, pondweeds, duckweeds, quillworts, watermeals, tapegrasses, and milfoils are common in aquatic beds. Submersed forms of emergent species may also be present. More than thirty rare species occur in

An **aquatic bed** community dominated by floating pondweed at Blow-Me-Down Pond in Cornish.

New Hampshire's aquatic bed communities. This is a very broadly defined community.

Good Examples of Drainage Marsh Natural Communities

SHRUB THICKETS AND SHRUBLANDS

Highbush blueberry - winterberry shrub thicket: Hopkinton-Everett Lakes Flood Control Area (Weare) and Grassy Pond (Litchfield).

Alder seepage thicket: Connecticut Lakes Natural Area (Pittsburg).

Alder alluvial shrubland: South River (Effingham).

Buttonbush shrubland: Stratham Hill Park (Stratham), Blackwater Flood Control Area (Salisbury), and Beaver Brook Association land (Hollis).

MEADOW MARSHES

Mixed tall graminoid - scrub-shrub marsh: Pawtuckaway State Park (Nottingham) and Watts Wildlife Sanctuary (Effingham).

Tall graminoid meadow marsh: Bear Brook State Park (Deerfield), Northwood Meadows State Park (Northwood), Pawtuckaway State Park (Nottingham), and South River (Effingham).

Short graminoid - forb meadow marsh/mudflat: Merrimack River (Concord), Dame Road vicinity (Durham), and Peverley Meadow (Canterbury).

Sedge meadow marsh: Pisgah State Park (Chesterfield), Deering Wildlife Sanctuary (Deering), Orenda-Stickey Wicket Wildlife Sanctuary (Marlow), Blackwater Flood Control Area (Salisbury), Norton Pool Preserve (Pittsburg), and Elbow Pond (Woodstock).

Herbaceous seepage marsh: College Woods (Durham), Great Bay National Wildlife Refuge (Newington), and Weeks State Park (Lancaster).

Lake sedge seepage marsh: Along the Pine River in Heath Pond Bog Natural Area (Effingham), Turtle Pond (Concord), Pickpocket Swamp vicinity (Exeter), and Great Bog (Portsmouth).

EMERGENT MARSHES AND AQUATIC BEDS

Emergent marsh: Ashuelot River (Lempster), Blackwater River (Salisbury), Exeter River (Exeter), Powwow River (Kingston), and around many ponds in the state.

Bayonet rush emergent marsh: Ossipee River (Freedom and Effingham) and Massabesic Lake (Auburn and Manchester).

Cattail marsh: Connecticut River (Hinsdale), Crommet Creek (Durham), Pawtuckaway State Park (Nottingham), and Smith Pond Bog (Hopkinton).

Aquatic bed: In most ponds and lakes in the state.

CHARACTERISTIC PLANTS OF SELECTED DRAINAGE MARSH NATURAL COMMUNITIES

A = Highbush blueberry - winterberry shrub thicket
B = Tall graminoid meadow marsh
C = Sedge meadow marsh
D = Cattail marsh
E = Emergent marsh
F = Aquatic bed

COMMON NAME	SCIENTIFIC NAME	A	B	C	D	E	F
SHRUBS & SAPLINGS							
Red maple	*Acer rubrum*	o		o	o		
Highbush blueberry	*Vaccinium corymbosum*	•		o	o		
Winterberry	*Ilex verticillata*	•		o	o		
Male berry	*Lyonia ligustrina*	o		o	o		
Leatherleaf	*Chamaedaphne calyculata*	o		o			
Speckled alder	*Alnus incana*	o		o	o		
Sweet gale	*Myrica gale*			o			
Meadowsweet	*Spiraea* spp.			o	o		
Poison sumac	*Toxicodendron vernix*			o			
HERBS							
Cinnamon fern	*Osmunda cinnamomea*	o		o			
Marsh fern	*Thelypteris palustris*	o	o	o	o		
Water horehound	*Lycopus uniflorus*	o	o	o	o		
Bluejoint	*Calamagrostis canadensis*		•	o	o		
Tussock sedge	*Carex stricta*		•	•	o		
Woolly bulrush	*Scirpus cyperinus*		•	o	o		
Reed canary grass	*Phalaris arundinacea*		•				
Cut-grasses	*Leersia* spp.		o	o	o		
Mannagrasses	*Glyceria* spp.		o	o	o		
Three-way sedge	*Dulichium arundinaceum*		o	o		o	
Joe-pye weeds	*Eupatorium* spp.		o	o	o		
Northern blue flag	*Iris versicolor*		o	o	o		
Canada rush	*Juncus canadensis*		o	o	o	o	
Silvery sedge	*Carex canescens*			•			
Bottle-shaped sedge	*Carex utriculata*			•			
Swamp candles	*Lysimachia terrestris*		o	o	o		
Lesser bur-reed	*Sparganium americanum*			o	o	•	
Common cattail	*Typha latifolia*			o	•	o	
Common arrowhead	*Sagittaria latifolia*					•	
Pickerel weed	*Pontederia cordata*					•	
Pondlilies & Waterlilies	*Nuphar* & *Nymphaea* spp.					o	•
Bladderworts	*Utricularia* spp.					o	•
Pondweeds	*Potamogeton* spp.					o	•
NON-VASCULAR							
Peat mosses	*Sphagnum* spp.			•			

• = abundant to dominant   o = occasional or locally abundant   * = state threatened or endangered

DRAINAGE MARSH NATURAL COMMUNITIES

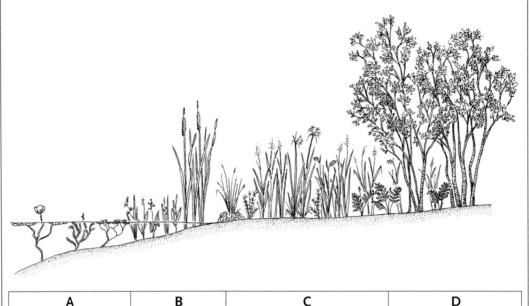

A	B	C	D

A. **Aquatic bed** – Floating-leaved and submersed plants in near-shore deep water

B. **Emergent marsh** – Herbaceous plants with soft, spongy tissue in shallow water

C. **Tall graminoid meadow marsh** – Grasses, sedges, rushes, and broad-leaved herbs in areas where summer water levels fall below the surface

D. **Alder alluvial shrubland** – Shrubs and herbs in seasonally flooded areas at the transition to swamp or upland forest

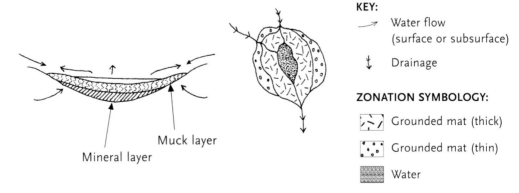

KEY:

→ Water flow (surface or subsurface)

↧ Drainage

ZONATION SYMBOLOGY:

Grounded mat (thick)

Grounded mat (thin)

Water

An idealized sequence of drainage marsh natural communities, with vertical plant scale greatly exaggerated relative to horizontal span. The smaller cross-section shows possible sources of water in these marshes (surface runoff, groundwater, and pond or stream overflow). The overhead view depicts stream inlets and outlets, and broad natural community zonation.

Sandy pond shore communities at Massabesic Lake in Auburn.

Sand Plain Basin and Sandy Pond Shore Marshes

Sand plain basin marshes and sandy pond shore marshes are distinct from other wetlands. Each is rare, has a globally restricted geographic distribution, and contains numerous rare and uncommon plants. In New Hampshire, these marsh types are concentrated on the sand plains of the lower Merrimack River Valley and Ossipee Lake region. Plants in sand plain basin and sandy pond shore marshes endure the stressful combination of a sandy, nutrient-poor substrate and widely fluctuating water levels. Those along pond shores are also subject to pronounced wave and ice disturbance. Sand plain basin and sandy pond shore marshes are concentrated in the Atlantic coastal plain region of North America.

Sand plain basin and sandy pond shore marshes share many of the same plants, but each also has distinct species and natural communities. Sand plain basin marshes occur in isolated, shallow depressions in sandy glacial deposits, and they lack stream inlets and outlets. The water table in the basins — whether perched above or connected to the local groundwater table — can fluctuate dramatically, varying as much as 7 feet in a single year. By contrast, sandy pond shore marshes occur around the shallow margins of larger, sandy-bottomed lakes where wave and ice disturbance is pronounced. Water levels fluctuate seasonally, but often less so than in basin marshes. In many instances, dams control or limit natural water-level variation.

Sand plain basin and sandy pond shore marshes support a tremendous number of plant species. Despite the rarity and small size of the communities, they collectively support more than 300 species, about one-seventh of all plant species in the state. More than half of these plants are short peren-

Sandy pond shore system at Ossipee Lake.

Grassleaf goldenrod is a rare coastal plain plant of sand plain basin and sandy pond shore marshes.

nials or annuals, and many have specialized growth forms or adaptations that confer an advantage under nutrient-poor, high-disturbance conditions. Other sand plain basin and sandy pond shore marsh plant species are habitat generalists that occur in a variety of wetland types. Sand plain basin and sandy pond shore marshes contain more than half (45 out of 87) of New Hampshire's coastal plain species, most of which are rare.

Plants and natural communities in sand plain basin and sandy pond shore marshes reflect sandy, nutrient-poor substrates, minimal input of external nutrients, wide water-level fluctuations and, on pond shores, significant wave disturbance. Herbaceous species are typically short and slow-growing. In basin settings, organic matter accumulation is low due to low biomass production and accelerated decomposition during frequent water drawdown periods. Organic deposits are thicker in permanently flooded zones. The location and variety of vegetation zones within a basin depends on its steepness and depth and the position of the water table. Seasonal and multi-year water-level fluctuations drive changes in the location and composition of vegetation. In pond shore settings, wave and ice disturbance erodes organic matter from lower and more exposed sections of shore and concentrates it in coves or sections of shore protected from prevailing winds, on the upper portions of berms, and in broad shallow areas. As a result, natural community distribution varies vertically and horizontally along the shore, reflecting complex gradients of disturbance, organic matter accumulation, and water levels.

In basin and pond shore settings, vegetation is a function of the response of buried seeds, called the seed bank, to seasonal and multi-year inundation and drawdown cycles. Many plants do not appear above ground in average or wet years, but emerge from the seed bank when water levels fall. Coastal plain plant species diversity is greatest in sandy or sandy-peaty zones subjected to regular or frequent drawdown.

The strategies and adaptations of basin and pond shore plants reflect stressful environmental conditions. The low nutrient levels and high disturbance in sand plain basin and sandy pond shore marshes favor annual plants and short, slow-growing perennials with low above-ground biomass. Water lobelia, for example, forms basal rosettes of short evergreen leaves and has specialized metabolic adaptations to contend with the low oxygen and nutrient conditions. The majority of sand plain basin and sandy pond shore plants rely on reproduction from seed rather than clonal growth, permitting persistence in a semi-dormant state during unfavorable, inundated growing conditions, and growth from seed when water levels recede. The seeds of many basin marsh and pond shore plants remain viable from decades to more than 100 years. The few clonal perennials of sand plain basin and sandy pond shore marshes tend to be short, and limited to drawdown zones between semi-permanently inundated areas and taller vegetation on drier ground adjacent to uplands. Annuals are common in intermittently exposed mudflats, where they grow from buried or recently wind-dispersed seed.

The main threat to New Hampshire's sand plain basin and sandy pond shore marshes is human activity. Altered surface and groundwater levels, dam construction and subsequently stabilized water levels, recreation, pollution, and development pose serious threats to these distinctive wetlands throughout their range. In addition to direct impacts to above-ground plants, many of these activities also reduce the diversity and abundance of seeds in the seed bank. Remarkably, invasive species are sparse at most sites, probably because exotic, invasive plants are not competitive with native plants under infertile, high-disturbance conditions.

Bulblet umbrella-sedge is frequent in sand plain basin and sandy pond shore marshes.

Sand Plain Basin Marshes

Basin marshes range from single, isolated basins to groups of basins in close proximity. Each basin differs in terms of the depth and steepness of the depression, water fluctuation, and number of natural communities. Shallow basins may contain a single community, whereas deeper basins may have as many as three or four communities. Highbush blueberry - winterberry shrub thicket, a drainage marsh community, may border basin marshes, or occupy the entire wetland area in some shallow basins.

Meadowsweet - robust graminoid sand plain marsh is restricted to the southeastern part of the state. The community occupies a position downslope of drier, tall shrub thickets and upslope of wetter, short herbaceous communities. A mix of robust perennial graminoids, shrubs, and aquatic peat mosses characterize this community. Meadowsweet dominates, with lesser amounts of tussock sedge and woolly bulrush. The vegetation is taller than

Meadowsweet - robust graminoid sand plain marsh at Grassy Pond in Litchfield, soon after a natural, late-summer, water-level drawdown.

Meadow beauty sand plain marsh at Grassy Pond in Litchfield.

other herbaceous basin marsh types, and the composition is reminiscent of a fen. The growth of peat mosses is prolific during wet periods, which can last for years, but decomposition during drawdown periods precludes the build-up of organic matter.

Meadow beauty sand plain marsh is a very rare community type of shallow, seasonally to semi-permanently flooded sand plain basins of south-central New Hampshire. Short clonal sedges, forbs, and clumped graminoids dominate over sandy, peaty soils. Abundant herbs include slender spike-rush, Virginia meadow beauty, lance-leaved violet, spurned panic grass, and spatulate-leaved sundew. Aquatic peat mosses are abundant and lush during wet periods, but reduce to a thin, dry layer during drawdown periods when herbs emerge from the seed bank. Meadow beauty sand plain marsh contains many species with coastal plain affinities.

Two marsh communities are restricted to closed basins of southern and central parts of New Hampshire. **Three-way sedge - mannagrass mudflat marsh** is a semi-permanently flooded community found in mucky sand plain basins. Periodic water-level drawdowns produce mudflat conditions with dense vegetation. Prominent life-forms include spongy-tissued, float-ing-stemmed, and other rhizomatous species, short clumped graminoids, and annual plants. Characteristic species include three-way sedge, swamp candles, several species of mannagrass, St. John's wort, and spike-rushes. Dominant plants in this community are broadly distributed species toler-ant of fluctuating water levels. Few if any coastal plain species are present. **Spike-rush - floating-leaved aquatic mudflat** occurs in wetter, lower areas than the three-way sedge - mannagrass mudflat marsh. It is intermittently exposed, and characterized by fewer tall graminoids than the previous com-munity, and more floating and submersed aquatic species. Perennial species

Montane sandy basin marsh
at Mount Stanton in Bartlett.

include various spike-rushes, mud rush, Torrey's threesquare, northern St. John's wort, northern floating mannagrass, golden pert, several pondweeds, and yellow pond lily.

Sharp-flowered mannagrass shallow peat marshes are semi-permanently flooded to intermittently exposed open, peaty swales. Sharp-flowered mannagrass and peat moss dominate. Other vegetation is sparse, but includes rattlesnake mannagrass, diverse-leaved water-starwort, various pondweeds, and yellow pond lily. Sharp-flowered mannagrass is a rare coastal plain species. The community is very rare in New Hampshire.

Montane sandy basin marsh is a rare community type found only in the White Mountains and North Country regions. This community occurs in areas with flat sand and gravel deposits at the bases of steep slopes. In most years, the water level drops below the surface of the marsh by the end of the growing season. In wet years, water may overtop the basin and flow into an ephemeral drainage stream. This community lacks the suite of coastal plain species found in more southerly basin marshes. Characteristic vegetation includes royal fern, Canada rush, mannagrass, water bulrush, meadowsweet, and northern blue flag. Species composition varies considerably among basins.

Sandy Pond Shore Marshes

Sandy pond shore marshes form linear bands along wave-washed and ice-scoured shores of permanent ponds and lakes. Communities can shift in a rapid sequence between the upland and the lake, and along the shore as well, as wave and ice disturbance decrease from exposed sections to more protected coves. Two of New Hampshire's sandy pond shore communities,

Sweet gale - speckled alder shrub thicket at Ossipee Lake.

Hudsonia inland beach strand near the mouth of the West Branch River on Ossipee Lake.

Twig-rush sandy turf pond shore at Lake Massasecum in Bradford.

hudsonia inland beach strand and twig-rush sandy turf pond shore, are globally rare.

Sweet gale - alder shrub thickets form a narrow zone at the upland edge of pond shores in southern and central New Hampshire. A diverse mixture of diagnostic tall and medium shrubs and ferns dominates, including sweet gale, speckled alder, steeplebush, witherod, and royal fern. Soils are highly variable, but typically consist of sandy organic mats over sand.

Hudsonia inland beach strand is a globally rare community that occurs on certain large lakes where shifting ice packs and wave action have created large, sandy berms on the shore. The berms remain dry and open, and are characterized by patches of wind-blown sand and scattered shrubs and herbs. The assemblage of plant species in this community includes hairy hudsonia, golden heather, scrub oak, little bluestem, ground juniper, dwarf cherry, pinweed, large cranberry, huckleberry, chokeberry, and various grasses.

Twig-rush sandy turf pond shore is also a globally rare community that occurs between open water and shrub communities along the sandy shores of larger lakes in southern and central New Hampshire. A mix of plant types, including tall clonal herbs and shorter, stress-tolerant graminoids and forbs, combines to form a moderately dense cover. Characteristic species include twig-rush, grass-leaved goldenrod, tussock sedge, wire sedge, bluejoint, lance-leaved violet, large cranberry and, occasionally, peat moss. Alternating periods of deposition and erosion produce turf soils that consist of variable mixes of sand and organic plant remains.

Bulblet umbrella-sedge open sandy pond shore is a sparsely vegetated community of wave- and ice-disturbed shores in central and southern parts of the state. The community forms a narrow strand between open water or unvegetated, wave-washed sand beach and more densely vegetated, less-disturbed areas on higher ground. The substrate is sand or sand with small quantities of organic matter. Short, clumped graminoids dominate, and rhizomatous forbs and graminoids are present. Native ruderals such as bulblet umbrella-sedge, lance-leaved violet, mud rush, and common beggar-ticks are present.

A sparse cover of floating-leaved and submersed aquatic plants characterizes **water lobelia aquatic sandy pond shore**.

The community occurs in intermittently exposed shallow-water environments subject to regular wave and ice disturbance. Common species include water lobelia, pipewort, several pondweeds, pickerel weed, and various reeds and rushes. One-flowered sclerolepis, a globally rare plant species, occurs in one example of this community.

Montane sandy pond shores occur in narrow zones along sandy pond shores in the White Mountains and northern New Hampshire. These wave-wracked shorelines support a mix of short, stress-tolerant marsh and fen plants. Species include northern bog clubmoss, several rushes, Pickering's reed bentgrass, spatulate-leaved sundew, prickly sedge, shore sedge, fertile yellow sedge, grass-leaved goldenrod, rough tickle grass, and a thin layer of disturbance-tolerant peat mosses.

Bulblet umbrella-sedge open sandy pond shore at Ossipee Lake.

Good Examples of Sand Plain Basin and Sandy Pond Shore Marshes

SAND PLAIN BASIN MARSHES

Meadowsweet - robust graminoid sand plain marsh: Grassy Pond (Litchfield).

Meadow beauty sand plain marsh: Grassy Pond (Litchfield).

Three-way sedge - mannagrass mudflat marsh: none on public lands.

Spike-rush - floating-leaved aquatic mudflat marsh: none on public lands.

Sharp-flowered mannagrass shallow peat marsh: none on public lands.

Montane sandy basin marsh: Bragdon Ledge (Albany) and Sugarloaf Mountain (Albany).

The rare sclerolepis occurs in one example of **water lobelia aquatic sandy pond shore** at Lake Massasecum in Bradford.

SANDY POND SHORE MARSHES

Sweet gale - alder shrub thicket: Ossipee Lake (Ossipee).

Twig-rush sandy turf pond shore: Lake Massasecum (Bradford) and Ossipee Lake (Ossipee).

Bulblet umbrella-sedge open sandy pond shore: Lake Massabesic (Manchester) and Ossipee Lake (Ossipee).

Water lobelia aquatic sandy pond shore: Lake Massabesic (Auburn) and White Lake State Park (Tamworth).

Hudsonia inland beach strand: Ossipee Lake (Ossipee).

Montane sandy pond shore: Gentian Pond (Success), Greeley Ponds (Livermore), Lily Pond (Livermore), and Umbagog Lake (Errol).

SANDY POND SHORE & SAND PLAIN BASIN MARSH NATURAL COMMUNITIES

SANDY POND SHORE MARSH NATURAL COMMUNITIES

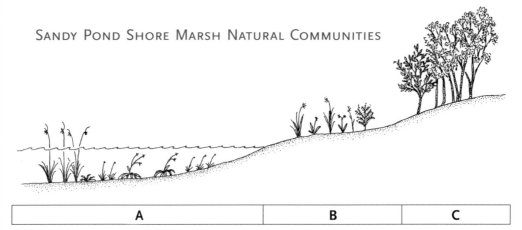

A	B	C

A. Water lobelia aquatic sandy pond shore – Semi-permanently inundated shallow water areas with rosette-leaved and other aquatic plants

B. Bulblet umbrella-sedge open sandy pond shore – Sparsely vegetated lake and pond shores subjected to wave action and ice-push disturbance

C. Sweet gale - alder shrub thicket – Dense, narrow shrub zone on berms and upland edges

SAND PLAIN BASIN MARSH NATURAL COMMUNITIES

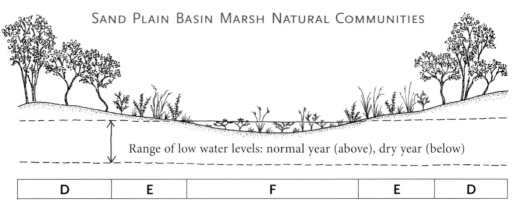

Range of low water levels: normal year (above), dry year (below)

D	E	F	E	D

D. Highbush blueberry - winterberry shrub thicket – Seasonally flooded border areas adjacent to swamp or forest

E. Three-way sedge - mannagrass mudflat marsh – Densely vegetated by short sedges, grasses, and forbs with spongy tissue; water levels drop below the surface in relatively dry years

F. Spike-rush - floating-leaved aquatic mudflat marsh – During dry years, the water table drops below surface and strands aquatic plants; annuals, short sedges, and forbs sprout from seed bank

CHARACTERISTIC PLANTS OF SELECTED SAND PLAIN BASIN MARSH AND SANDY POND SHORE MARSH NATURAL COMMUNITIES

A = Meadowsweet - robust graminoid sand plain marsh
B = Three-way sedge - mannagrass mud flat marsh
C = Spike-rush - floating-leaved aquatic mud flat marsh
D = Sweet gale - alder shrub thicket
E = Twig-rush sandy turf pond shore
F = Water lobelia aquatic sandy pond shore

COMMON NAME	SCIENTIFIC NAME	A	B	C	D	E	F
SHRUBS & SAPLINGS							
Meadowsweet	Spiraea alba	•					
Leatherleaf	Chamaedaphne calyculata	o					
Large cranberry	Vaccinium macrocarpon	o			o		
Sweet gale	Myrica gale				•	o	
Speckled alder	Alnus incana				•		
Steeplebush	Spiraea tomentosa				o		
Viburnums	Viburnum spp.				o		
HERBS							
Woolly bulrush	Scirpus cyperinus	o					
Three-way sedge	Dulichium arundinaceum	o	•			o	
Mannagrasses	Glyceria spp.	o	•	o		o	
Swamp candles	Lysimachia terrestris	o	•				
Tussock sedge	Carex stricta	o			o	o	
Spike-rushes	Eleocharis spp.		o	•		o	
St. John's-worts	Hypericum spp.		o	o			
Lance-leaved violet	Viola lanceolata		o			o	
Bottle-shaped sedge	Carex utriculata		o				
Yellow pondlily	Nuphar variegata			•			
Pondweeds	Potamogeton spp.			o			•
Bladderworts	Utricularia spp.			o			o
Royal fern	Osmunda regalis				o		
Grassleaf goldenrod*	Euthamia caroliniana*					o	
Flat-topped goldenrod	Euthamia graminifolia				o	•	
Twig-rush	Cladium mariscoides					•	
Wire sedge	Carex lasiocarpa					o	
Bluejoint	Calamagrostis canadensis					o	
Water lobelia	Lobelia dortmanna						•
Pipewort	Eriocaulon aquaticum						•
Pickerel weed	Pontederia cordata						•
Arrowheads	Sagittaria spp.						o
Lesser bur-reed	Sparganium americanum						o
Sclerolepis*	Sclerolepis uniflora*						o
NON-VASCULAR							
Peat mosses	Sphagnum spp.	•	o				

• = abundant to dominant   o = occasional or locally abundant   * = state threatened or endangered

Marsh Wildlife

Acre for acre, marshes have a higher concentration and diversity of wildlife species than almost any other habitat. In drainage marshes productive wetland soils, dynamic nutrient cycles, and emergent vegetation provide food, nesting, shelter, and rearing habitats for a variety of common and rare wildlife. Small emergent plants provide shelter for newly hatched sunfish, bass, and brown bullheads in the shallows of some sandy pond shores. Moose, beaver, muskrat, and mink are some of the mammals that spend all or part of their life cycle in or near marshes. Water depth and marsh area largely determine which wildlife species inhabit a particular marsh. For example, American bitterns, ribbon snakes, green frogs, and sedge wrens prefer shallow-water areas. In contrast, deeper water attracts snapping turtles, musk turtles, sunfish, common moorhens, and great blue herons.

Dragonflies, mayflies, caddis flies, midges, and mosquitoes spend their juvenile stages under water, and often feed over the water or in marsh vegetation as adults. Dragonflies and damselflies perch and lay their eggs on the stems of grasses, rushes, and sedges. The larvae of the common sanddragon (a dragonfly) burrow into the sand for protection from predators, and wait there in hiding for smaller insect prey to approach. The rare scarlet bluet, a small red damselfly, perches on floating leaves in southern marshes.

These wetlands are critical habitat for many amphibians and reptiles, including ribbon and northern water snakes, bullfrogs, green frogs, and painted, musk, spotted and snapping turtles. Mink frogs are limited to certain emergent marshes and aquatic beds in the northern part of the state. They bask on water lilies, where they lay in wait for spiders, snails, beetles, and other invertebrate prey.

Red-winged blackbirds, marsh and sedge wrens, and yellow warblers nest in shrubs and cattails. Tree swallows nest in cavities of dead standing trees and feed on marsh insects. Black ducks and common moorhens forage for plants and small invertebrates in the water and mud. Other waterfowl use marshes as stopover feeding and resting locations during migration. Great blue herons build stick nests in tall, dead trees around marsh edges. Proximity to the marsh provides the herons with easy access to frogs and insects for food. Great horned owls may use heron nests in late winter.

7 River Channels and Floodplains

Thousands of miles of rivers and streams meander throughout the New Hampshire landscape. These narrow corridors drain water from mountain and hills, pass through forests and fields, and connect lakes and ponds to the sea. The lands adjacent to and directly influenced by rivers consist of river channel and floodplain environments, collectively called the riparian zone. While rivers and their riparian zones occupy less than 2 percent of the state's total area, they nevertheless are rich in plants and animals and contribute greatly to its ecological diversity.

Riparian communities consist of two broad categories: river channels and floodplains. A river channel is the area between the tops of the riverbanks (excluding the river itself). Floodplains are flat terraces beyond the tops of riverbanks, and are inundated by occasional floods. The frequency and duration of inundation in riparian zones depends on water volume and differences in elevation. For example, low channel areas are frequently inundated for long periods during spring snowmelt and other high-water events, whereas the sloped riverbanks are inundated less frequently and for shorter periods. The point at which a river channel is full of water, just to the top of its banks, is called the bankfull stage. Flows higher than bankfull result in flooding, starting with the low floodplain. Floods occur every year or two on most rivers. Steep or deeply entrenched sections of rivers or streams may have little or no floodplain development. Floods in these sections may never reach the top of the riverbank. In both river channel and floodplain settings, variation in the frequency, intensity, duration, and timing of floods determines species composition and natural community expression.

River gradient or slope, flow regime, and sediment size affect the patterns and features of the riparian zone. In general, as river gradient, water volume, or velocity increase, so do the sediment volume and particle size being transported. The energy of moving water alters the river channel, entrenching or widening the channel, eroding riverbanks and bars, and depositing

PRECEDING PAGE: Riparian communities along the Saco River in Conway.

Spring flooding along the Soucook River in Loudon.

High-gradient section of Wildcat Brook in Jackson.

sediments that expand or form new bars and floodplains. Most of this work occurs when flows are at or above bankfull, when velocity and volume are highest. As the river gradient decreases and velocity diminishes, sediments of progressively smaller size fall out of suspension and form deposits on the channel bottom. Elevation typically does not decrease by continuous decline along a river, but rather by an alternating series of high- and low-gradient sections. Steep drops in elevation, such as waterfalls or rapids, are closest to each other on high-gradient rivers, and occur farther apart on meandering, low-gradient rivers.

The distribution of riparian natural communities is closely associated with river gradient, sediment size, and other channel characteristics. River gradient is generally higher in headwater areas and lessens downstream. High-gradient sections of rivers are common in the upper portions of watersheds. They contain high-energy channel communities, but rarely support floodplains. These steep, fast rivers transport fine sediments, leaving boulders or bedrock in the channel. High-gradient river segments occur in narrow valleys, entrench rather than widen, and form relatively straight, stable channels. Steps and pools are often closely spaced.

Moderate-gradient river sections are most common in middle watershed positions. They support more channel communities, and sometimes have floodplain communities. These sections can have stable or unstable banks, with channels that are somewhat more sinuous than in high-gradient sections. They can be entrenched with little or no floodplain or have only minor floodplains. Steps and pools are more widely spaced than on high-gradient river sections. Rivers with flashy flood regimes—those with extreme annual variation in flow, including intense, short-duration floods—are less en-

Moderate-gradient section of the Pemigewasset River in Plymouth.

trenched and form dynamic, braided channels with lots of riffles, sandbars, and islands. Such features are common on moderate- to low-gradient rivers with heavy sediment loads.

Low-gradient river sections are most common in lower watershed positions. They have relatively extensive forested floodplain and high riverbank

Low-gradient section of the Merrimack River with well-developed meanders, floodplains, and high terraces.

communities, while low channel communities are scarce or absent. In these slower, flatter sections, the forces of erosion and deposition are in relative balance. Rivers form sinuous, meandering channels bordered by floodplains with sand or silt soils, often in broad valley bottoms. The breadth of the meanders and the distance between them generally increases with the volume of flow. Some deeply entrenched low-gradient rivers have no low channel communities, but feature well-developed riverbank communities. Because the terraces are too high to flood, no floodplain forests occur.

Flood pulses of varying frequencies, intensities, and durations create complex mosaics of natural communities in riparian zones. Collectively, riparian natural communities have many plant species exhibiting diverse life-history strategies. Channels have fewer trees than floodplains, and more shrubs and herbs adapted to longer periods of inundation, more intense disturbance, shifting sediments, and lower nutrient levels. Because of these dynamic channel conditions, the species composition of channel communities is more variable than in floodplain communities, and community locations can shift as channel features change. Floodplains are dominated by trees, vines, and robust, competitive herbs that grow fast and tolerate or benefit from short periods of inundation, moderate disturbance levels, and relatively high nutrient levels.

More than a third of New Hampshire's vascular plant species occur in

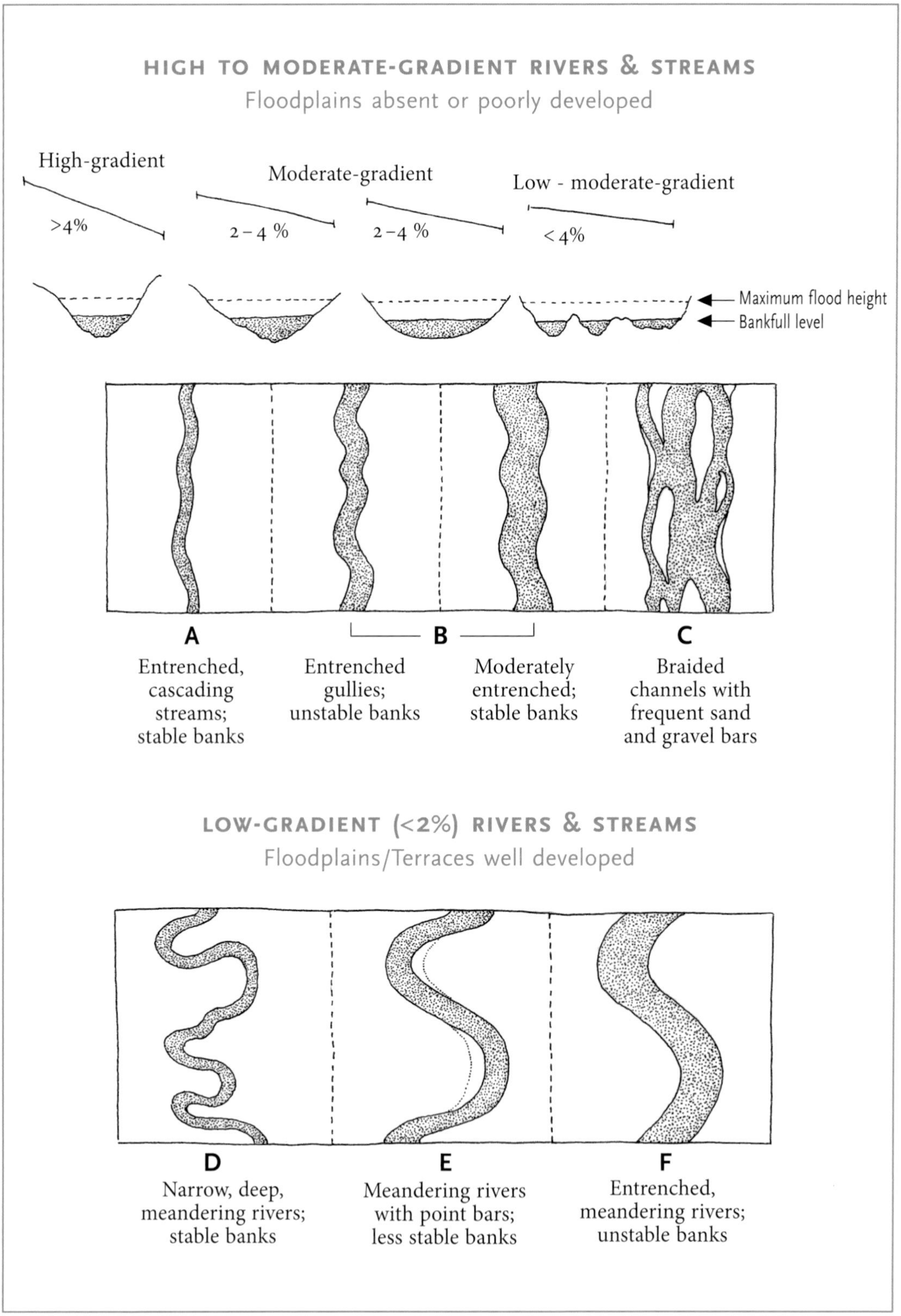

HIGH TO MODERATE-GRADIENT RIVERS & STREAMS
Floodplains absent or poorly developed

High-gradient

Moderate-gradient

Low - moderate-gradient

>4%

2 – 4 %

2 – 4 %

< 4%

← Maximum flood height
← Bankfull level

A

Entrenched,
cascading
streams;
stable banks

B

Entrenched
gullies;
unstable banks

Moderately
entrenched;
stable banks

C

Braided
channels with
frequent sand
and gravel bars

LOW-GRADIENT (<2%) RIVERS & STREAMS
Floodplains/Terraces well developed

D

Narrow, deep,
meandering rivers;
stable banks

E

Meandering rivers
with point bars;
less stable banks

F

Entrenched,
meandering rivers;
unstable banks

RIVERS IN NEW HAMPSHIRE

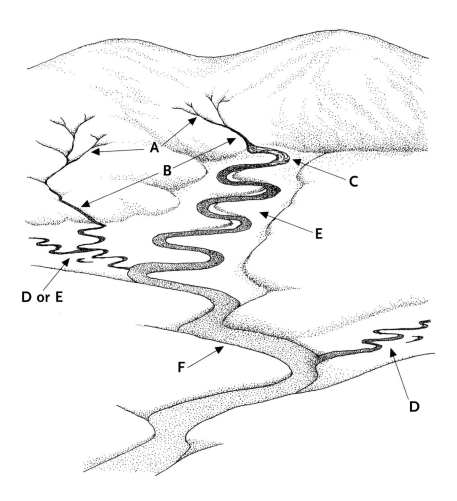

River patterns and processes change along the course of a river. The diagrams on this and the facing page show some typical combinations of river features along high, moderate, and low-gradient sections of river with varying degrees of floodplain or channel feature development. Features illustrated include degree of entrenchment, flood-prone area, bank stability, channel pattern, and meander development. The locations of these idealized types of rivers and streams are depicted in a hypothetical landscape, intended to simulate the Pemigewasset and Merrimack river valleys. Although high-gradient areas are most common high in a watershed, and low-gradient areas most common low in a watershed, changes in gradient and river type can occur anywhere along the course of a river. Adapted from Rosgen (1996).

1. LARGE, LOW-GRADIENT RIVER

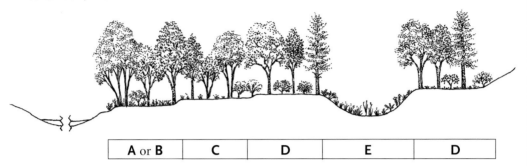

A or B	C	D	E	D

2. MODERATELY LARGE, MODERATE TO LOW-GRADIENT FLASHY RIVER

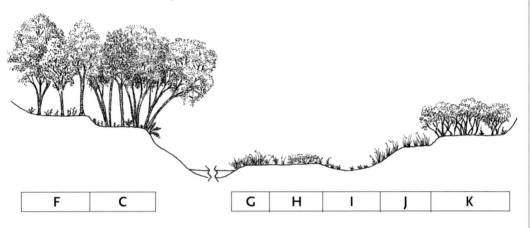

F	C		G	H	I	J	K

3. SMALL, LOW-GRADIENT RIVER OR LARGE STREAM

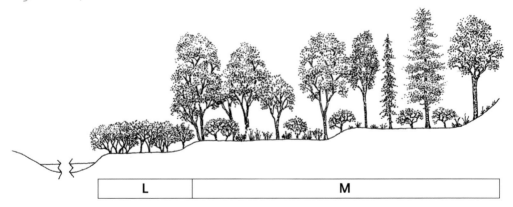

L	M

Idealized sequences of river channel and floodplain natural communities on three rivers. Vertical scale of plants is greatly exaggerated relative to horizontal span. 1. Large, low-gradient river (more than 250 ft. wide at low water) along the main stems of the Connecticut or Merrimack rivers. 2. Moderately large, moderate to low-gradient flashy river (50–250 ft. wide at low water). The channel contains a heavy sediment load and well-developed sand and gravel bars and channel shelves. 3. Small, low-gradient river (25–50+ ft. wide) with few channel bars or shelves.

RIVER CHANNEL AND FLOODPLAIN NATURAL COMMUNITIES

A. **Silver maple - wood nettle - ostrich fern floodplain forest**
Low floodplains along the Connecticut River

B. **Silver maple - false nettle - sensitive fern floodplain forest**
Low floodplains on large rivers, including major tributaries of the Merrimack, Ashuelot, and Contoocook rivers

C. **Sugar maple - silver maple - white ash floodplain forest**
On middle terrace floodplains of large rivers or low floodplains of large, flashy rivers

D. **Upland forest on high terrace**
Various communities are possible, such as **rich mesic forest, semi-rich oak - sugar maple forest,** or **hemlock - beech - oak - pine forest**

E. **Emergent marsh**
Herbaceous marsh plants in abandoned river channels (oxbows); any combination of emergent marsh, meadow marsh, and shrub communities can occupy an oxbow channel, depending on the depth of the basin and fluctuations of water

F. **Sugar maple - ironwood - short husk floodplain forest**
Moderate to high floodplains along flashy rivers near the White Mountains

G. **Twisted sedge low riverbank**
Large sedge tussocks on channel bars or low riverbanks of flashy low to high-gradient rivers

H. **Dwarf cherry river channel**
Sand and gravel bars on moderately large rivers with heavy sediment loads

I. **Herbaceous river channel**
Grasses, sedges, and other herbs on sandy or cobbly low channel substrates exposed at low water

J. **Bluejoint - goldenrod - virgin's bower riverbank/floodplain**
High riverbanks and adjacent floodplains along flashy rivers

K. **Alder alluvial shrubland**
Temporarily flooded high channels and low floodplains on small or large rivers

L. **Alder - dogwood - arrowwood alluvial thicket**
Diverse shrub thickets on low floodplains of small rivers

M. **Red maple floodplain forest**
The most common type of floodplain forest on small rivers; other types are possible on northern and coastal rivers

riparian natural communities, including ninety-three rare species. Most of the rare plants grow in open river-channel communities, consistent with a general ecological pattern of higher rare-plant diversity in disturbed or stressful habitats. Riparian communities also support more invasive plants than any other major group of natural communities. Their productive, routinely disturbed soils, and proximity to roads, agriculture, and development, promote invasive plant colonization and growth.

The integrity of riparian natural communities depends on the maintenance of natural flood regimes. Dams, land development, and water consumption cause changes that affect the distribution and abundance of plants, birds, and insects adapted to natural flood pulses. Dams in particular cause significant changes to riparian environments. They trap sediments that would normally replenish downstream floodplains, sometimes burying and eliminating historic floodplain wetlands that existed upstream of the dam. Dams that eliminate major floods — those floods with intervals of 10 or more years — diminish or alter high-terrace floodplain communities downstream. Alterations to the timing and frequency of flood disturbances and other physical conditions on floodplains shift characteristic plant communities by affecting seed germination and growth, altering competition between species, and facilitating the spread of invasive species. Encouraging the return of more natural flood regimes is a critical component of riparian-area conservation efforts.

Invasive Species

Many conservationists consider invasive plant and animal species the second most critical threat to the persistence of native biodiversity (habitat loss is first). A non-native species is one that exists outside its native range, having arrived into a new region as a result of either intentional or unintentional human-related activities. Intentional introductions of non-natives include those chosen for use in horticulture or landscaping; plants and animals that "hitchhike" on packing crates or other imported products are examples of unintentional introductions. Invasive plants are a subset of non-native species whose introduction to an ecosystem has the potential to cause environmental harm. Once established, invasive plants produce many viable seeds or other propagules that travel by wind and water, animal transport, or inadvertently through human activity.

Disturbed areas, both human and natural, are especially susceptible to the establishment and spread of invasive species. In riparian areas, combinations of frequent ice and water scour, rich floodplain soils, increased sunlight, and water-transport dispersal opportunities provide excellent conditions for these invaders to establish, grow, and spread.

Some of the more troubling invasive plant species occurring within riparian zones include Japanese knotweed, glossy buckthorn, and honeysuckle. These species shade out natives, lower native plant diversity, and reduce habitat value.

The threat of invasive species is considerable in New Hampshire. Best-management practices require multiple approaches including inventory and mapping, control at the most ecologically important or vulnerable sites, policy and legislation, and regional partnerships.

Japanese knotweed, an invasive plant, along the Baker River in Wentworth.

River channel communities along the Pemigewasset River.

River Channels

River channels are dynamic, frequently flooded environments below the tops of riverbanks. During periods of significant snowmelt or heavy rain events, rivers swell and overtop their banks. Most of the time, however, rivers remain within the confines of their channels. Water rises and falls many times during the course of a year, revealing vegetation along sloped riverbanks and on bars and shelves lower in the channel. Flowing water continuously erodes and deposits sediments and plant debris within the channel environment.

A river's flood regime is the frequency, intensity, duration, and timing of its floods. Flood regime varies with river gradient and location within the channel. High-gradient rivers occur high in a watershed and can have frequent, high-energy, and short-duration floods at any time of year. Low-gradient rivers occur lower in the watershed and have fewer, less intense, and longer-duration floods, primarily during spring and fall. Low channel settings are inundated for substantial portions of the year, regardless of gradient. Low channels include sand or cobble bars and shelves, and adjacent low riverbanks. By contrast, high channels (including high riverbanks) flood less frequently, but the intensity can be high here along many rivers, particularly if the riverbank is regularly scoured by ice.

River gradient, sediment size, and channel location are therefore good predictors of the channel communities. Communities of high- to moderate-gradient sections of rivers, and those in low channel locations, have coarse rock or cobble substrates, little accumulated organic matter, and sparse plant cover. In contrast, communities of low-gradient river sections and higher channel settings have finer sand and silt substrates, more accumulated organic matter, and denser plant cover. Channel communities generally exist

The Peabody River north of Pinkham Notch.

During floods, smaller sediment particles settle out in lower-velocity areas of river channels, including the downstream side of rooted plants.

as narrow bands, and change rapidly with small differences in elevation. Stagnant areas along some low-gradient rivers, such as muck-bottomed backwater channels, support emergent marsh and aquatic bed communities (see chapter 6).

Perennial plants growing in river channels tolerate periodic inundation. Flooding saturates the soil, inhibiting plant oxygen uptake and respiration. Emergent and aquatic plants thrive in semi-permanently to permanently inundated areas, however. These species are structurally flexible, have soft, spongy tissue, and absorb oxygen directly from the surrounding water. By contrast, plants growing in drier channel settings, including high on river-banks, need only endure brief periods of flooding and ice scouring. Adaptive strategies that help woody plants survive frequent inundation and wet soils include buttressed bases, multiple stems that increase surface area and oxygen intake, and adventitious, above-ground roots. Some riparian plants also have the ability to decrease flowering during floods and increase flowering during dry periods.

Plants in the river channel are also adapted to seasonal drought and annual scouring from ice and strong river currents. Stress-tolerant channel plants include dwarf shrubs, grasses, sedges, composites, and certain wetland herbs. Plants in some channel settings endure inundation in the spring and early summer, and drier conditions by mid to late summer. Many perennials in these settings have deep roots that can access water during drought periods. Weedy annual forbs and graminoids that excel in disturbed environments are abundant in channel settings. They avoid periods of inundation as short-lived seeds buried in river sediments, and rely on rapid annual growth and seed production for perpetuation.

A few shrubs and trees are specially adapted to the scour and burial stresses typical of channel environments. Willows and alders have abrasion-resistant, narrow, and flexible stems. Dwarf cherry and some species of willow sprout roots from stems that grow along the ground. They also send shoots to the surface when buried by sand or gravel. Certain willows are well adapted to disturbed, low channel-bar environments exposed after spring's high waters recede. Willow seeds are viable for only short periods, and require wet, open soil conditions to germinate. Willow species disperse their seeds at different times during the growing season, producing species bands at progressive elevations on channel bars and banks. Alder has the ability—with the aid of microorganisms—to procure or "fix" nitrogen directly from air in the soil. The nitrogen-fixing process confers a distinct advantage in river-channel environments, as it does in other nutrient-poor, early successional habitats. Riverbanks and adjacent floodplain areas are less disturbed and have soils with more organic matter and fine sediments. These conditions favor fast-growing, competitive, and flood-tolerant plants.

A dense cover of robust herbs, vines, shrubs, and tree saplings dominates these areas.

River channel natural communities are differentiated by channel location, substrate, summer moisture level, and dominant life-forms. Hydric to mesic low channel natural communities on low riverbanks, channel shelves, and bars are inundated in spring and early summer but exposed by midsummer. Soils retain water near the surface through the growing season. Dry low channel communities are wet early in the growing season, but surface conditions are usually dry by midsummer. High river channel communities are flooded less frequently and for shorter durations, and occur in medium to high riverbank and adjacent floodplain settings. Riverbank outcrops and seeps occur mostly along river narrows, and extend from low to high riverbanks. Many river channel communities migrate over time with shifts in the river's course. As with other communities that experience significant disturbance, species composition can vary from one example to another.

Hydric to Mesic Low Channels

Four low channel communities remain wet or moist through most of the growing season. They are most common on flashy, low- to moderate-gradient sections of rivers and large streams with heavy sediment loads.

Two herb-dominated communities occur in coarse substrate, low channel settings along rivers and large streams. Shrubs are absent or sparse in each. **Mesic herbaceous river channels** form on sand and cobble bars and channel shelves in moderate- to high-energy settings. Moisture conditions are mesic to hydric through most or all of the growing season. Short to moderate-height herbaceous wetland forbs, grasses, sedges, and ferns comprise a sparse to dense cover. Cardinal flower, water purslane, and beggar-ticks are common. Plants that survive in emerged or submerged conditions, such as smartweeds and swamp candles, are also present. In some examples of this community, higher portions dry out in late summer and support many of the same short, stress-tolerant graminoids found in sandy pond shore marshes, including cut-grasses, rushes, and sedges such as bulrushes and spike-rushes. **Twisted sedge low riverbank** occurs as a narrow zone at the base of some high- to moderate-energy riverbanks on cobbly, gravelly, or sandy material near the river's edge, or as

Mesic herbaceous river channel along the Pemigewasset River at Livermore Falls.

Twisted sedge low riverbank along the Peabody River south of Gorham.

Willow low riverbank along the Piscataquog River in New Boston.

more extensive areas on low bars. Twisted sedge is always present and forms large green tussocks or patches, with other species sometimes present in low abundance.

Willow low riverbanks occur slightly higher above the river than the previous two communities, and have fewer graminoids. They support a sparse to moderate cover of willow, and lesser amounts of alder, other shrubs, tree saplings, woody vines, and herbs. This community is typically narrow and discontinuous, sometimes featuring bands of willow seedlings that establish as falling water levels expose sediments.

The **riverweed river rapid** community forms low mats on submerged rocks in rapids, waterfalls, and other fast-flowing sections of streams and rivers in central and southeastern New Hampshire. It usually occurs as a riverweed monoculture, although pondweeds may grow in gravel deposits between riverweed-covered rocks. Riverweed is a seed-bearing plant that adheres to rocks using sucker-like roots. Seed-bearing plants are otherwise uncommon in permanently flooded, high-energy channels.

Dry Low Channels

Five river channel communities are inundated at high water and dry by mid-summer. River gradient, channel material, and vegetative cover and composition distinguish these communities, which are most abundant on flashy rivers or streams with heavy sediment loads. One community occurs along high-gradient rivers and streams, the other four along moderate- to low-gradient rivers.

Boulder - cobble river channels are common along flashy, high-gradient streams and small to medium-sized rivers in the White Mountains and hilly parts of the state. This community floods rapidly and intensely with snowmelt runoff in the spring, and also during peak rain events. As a result, most woody vegetation is short, and the community is sparsely vegetated by alders, willows, and other shrubs, tree seedlings, grasses, and composites. Higher areas dry out in the summer, whereas areas closer to the river's edge remain moist. Rare subalpine plants found in some examples of this community in or near the White Mountains include mountain avens and New England northern reedgrass.

Cobble - sand river channels occur on sandy, gravelly, and cobbly point-bars, shelves, and islands on low- to moderate-gradient rivers with high sediment loads. The community is sparsely vegetated, but species richness

Boulder - cobble river channel along the Peabody River.

253

Cobble - sand river channel along the Connecticut River in Lebanon.

Hudsonia - silverling river channel along the Saco River in Conway.

Dwarf cherry river channel along the Pemigewasset River at Livermore Falls in Holderness.

can be quite high in some examples. Sedges, grasses, composites, and forbs are the dominant herbs. Scattered dwarf shrubs and tree seedlings may also be present. The state-threatened cobblestone tiger beetle occurs in some examples of this community along the Connecticut River.

In New Hampshire, the globally rare **hudsonia - silverling river channel** community occurs only along the Saco River. The community forms narrow strips of vegetation on channel shelves, just above the river and below the forest edge. Grasses, forbs, mosses, and lichens are the dominant life-forms, although vegetation is sparse overall. The association of two rare perennial species, hairy hudsonia and silverling, is otherwise known only from ridgetop barrens on Panther Knob in West Virginia. Elsewhere in New England, hairy hudsonia occurs on inland beach strands and coastal dunes, and silverling grows on outcrops and mountain cliffs.

Dwarf cherry river channel communities occur on exposed channel bars and shelves of sand, gravel, or cobble along the Connecticut and Pemigewasset rivers. Dwarf cherry is the dominant plant. Grasses, goldenrods, and other plants are usually present, including little and big bluestems (native prairie grasses), and upland bentgrass. Following spring floods, this community is dry or moist for most of the growing season. Dwarf cherry forms clonal patches, sometimes trapping organic debris and creating fluvial sand dunes downstream.

Mixed alluvial shrublands occur on sandy to cobbly channel shelves above low riverbanks or bar communities and below high riverbank or floodplain communities. They are highly dynamic and shift with point-bar migration. Tree saplings and shrubs total more than 25 percent cover, and herbs are relatively sparse. Invasive plant species are common and occasionally abundant in this natural community.

Herbaceous High Channels and Floodplains

Herbaceous communities in high channel settings and adjacent floodplains form a transition zone between low river channel communities and forested floodplain or upland forests. In most of these communities, vegetation is dense and consists of herbs and sparse shrubs and trees. Soils vary with slope and proximity to the main channel. Consolidated sandy and loamy soils are typical, but gravel and bedrock occur in high-energy settings and silt occurs in low-energy settings. These communities occur along low- to moderate-gradient sections of nearly all large and small rivers away from the mountains.

Herbs dominate two medium to high riverbank communities, both of which can extend onto adjacent floodplain settings. These communities can be similar in appearance to meadow marshes. In contrast to meadow marshes, however, there is rapid decomposition during periods of low water and little or no organic matter accumulation due to flood and ice scour. **Herbaceous riverbank/floodplains** are wet to mesic meadows on silty or sandy soils along the banks of large and small rivers, and large streams. They are dominated by tall herbs, including reed canary grass, big bluestem, goldenrods, robust ferns, and various sedges. Shrubs and tree saplings mix with

Mixed alluvial shrubland along the Pemigewasset River at Livermore Falls in Holderness.

Herbaceous riverbank/ floodplain along the Merrimack River in Canterbury.

Bluejoint - goldenrod - virgin's bower riverbank/floodplain along Indian Stream in Pittsburg.

herbs in some examples, but total less than 25 percent cover. **Bluejoint - goldenrod - virgin's bower riverbank/floodplains** are communities that occur along flashy, small rivers and large streams. Bluejoint, rough goldenrod, smooth goldenrod, and virgin's bower dominate, with lesser amounts of other tall herbs and vines. This community occurs on medium-grained sands between low riverbanks and shrubby or forested floodplains.

Two communities form in unusual riparian settings with deep lakebed, outwash, or deltaic sand deposits along the lower Merrimack and Saco river valleys. **Dry river bluffs** occur on tall, sloughing, sandy, outside bends of river meanders. Depending on past disturbance, these communities can range from open, sparsely vegetated areas to woodlands. The river erodes the bottom of the bluffs, causing sand to slide down the steep banks. Vegetation takes root as slide areas stabilize. Clumps of intact vegetation detach

Dry river bluff along the Merrimack River in Canterbury.

Riverwash plain and dunes on a broad terrace along the Merrimack River in Canterbury.

from the top of the bluff and either settle downslope or slide into the river. Dry-site grasses, sedges, and forbs including little bluestem, big bluestem, Allen's poverty oatgrass, awnless brome grass, and a variety of sedges and mosses are common. Paper birch occurs less frequently. Two rare plants—golden heather and wild lupine—are sometimes present.

Inland dunes are a rare geologic phenomenon in New Hampshire, and only occur in the globally rare **riverwash plain and dunes** on the Merrimack River in Canterbury. Here, this community is located on an inside river bend's high floodplain terrace. Vegetation is characterized by a sparse to moderate cover of drought-tolerant grasses, sedges, and forbs, along with a fragile cryptogamic crust. Cryptogamic crusts are ecologically important biotic layers consisting of some combination of mosses, lichens, fungi, bacteria, and algae. For thousands of years, floods—and occasional fires—have maintained the open conditions of this community. Major flood events periodically inundate the terraces, scouring and burying portions with fresh sand and gravel. Winds subsequently rework the sand deposits to form low dunes, up to 5 feet high. The dunes range from open to semi-wooded.

Shrub High Channels and Floodplains

Shrubs dominate three riverbank communities common along low- to moderate-gradient sections of streams and rivers. The communities often flank herbaceous riverbank and upland forest communities. **Alder alluvial shrublands**, also associated with drainage marshes, occur on riverbanks and floodplains in low- to moderate-energy settings. They are common in the northern part of the state, and more sporadic south of the mountains. Speckled alder dominates the shrub layer, and the herbaceous layer is variable in density and composition. Shrub stems lean in the direction of river

Alder alluvial shrubland along Indian Stream in Pittsburg.

Alder - dogwood - arrowwood alluvial thicket at St. Gaudens National Historic Site in Cornish.

Meadowsweet alluvial thicket along the Soucook River in Loudon.

Acidic riverbank outcrop along the Merrimack River in Manchester.

flow and exhibit signs of flood scour on the upstream sides of their stems. These shrublands flood every year or two. **Alder - dogwood - arrowwood alluvial thicket** is a similar community common in central and southern parts of the state. It frequently has a dense mix of several shrub species, though trees, woody vines, and herbs may be present. This community includes examples dominated by dogwoods, such as red osier dogwood, with few other shrub species. The **meadowsweet alluvial thicket** community occurs on large streams and small rivers throughout the state. As the name suggests, meadowsweet is the dominant plant, although other woody species may be present. Robust herbs such as bluejoint, sallow sedge, and swamp candles may also be present.

Riverbank Outcrops and Riverside Seeps

Riverbank outcrops are scattered around the state on medium- to large-sized rivers, but are most common along the middle third of the New Hampshire portion of the Connecticut River. They commonly occupy low to high riverbank positions in narrows, and occasionally occupy wider stretches, where ice and water scour expose bedrock. Vegetative cover is sparse and includes both flood- and drought-tolerant species. Riverbank outcrop communities occur on both acidic and circumneutral substrates.

Acidic riverbank outcrops are open, flood-scoured bedrock exposures along medium- to large-sized rivers. Vegetation is limited to cracks in the bedrock where acidic soils accumulate. Plants include a variety of forbs, grasses, and ferns, such as pointed auricle path rush, Allen's poverty oatgrass, hawkweeds, goldenrods, and northern lady fern. Lichens and woody plants are notably absent. Less common than their acidic counterparts, **circum-**

neutral riverbank outcrops occupy similar settings, but with substrates that have a neutral pH due to higher bedrock calcium levels. Plants associated with these conditions, and absent on acidic outcrops, include harebell, rusty woodsia, Christmas fern, and rare plants such as dwarf ragwort, Siberian chives, and the globally rare Jesup's milk vetch. In New Hampshire, the latter community is restricted to two sites in the Connecticut River Valley, and it is rare elsewhere in New England.

Riverside seeps occur where cold groundwater emerges from bedrock fractures and spaces between cobble, sand, or silt substrates of flood-scoured shores along larger rivers. Such settings often occur at natural constrictions below dams built at river narrows. Riverside seeps share some similarities with fens, and support many of the same species. Floodwaters and ice floes scour and sweep away woody plants, removing potential competitors to herbaceous species. Characteristic species of **acidic riverside seeps** include leatherleaf, large cranberry, large-leaved sundew, and small-flowered gerardia. This community is rare in New England, and occurs in New Hampshire along the Connecticut, Ammonoosuc, Pemigewasset, and Merrimack rivers. Most examples occur on seepy outcrops. **Calcareous riverside seeps** occur on turfy, sandy substrates wedged in bedrock cracks or in spaces between cobbles or boulders. Forbs, grasses, sedges, and some Midwestern prairie species like big bluestem are the dominant life-forms. A number of rare plants differentiate this community from its acidic counterpart, including Kalm's lobelia, sticky false asphodel, grass-of-parnassus, and Garber's sedge. Calcareous riverside seeps occur only in the Connecticut River Valley and are often associated with circumneutral riverbank outcrops.

Circumneutral riverbank outcrop at Sumners Falls on the Connecticut River.

Calcareous riverside seep at Sumners Falls on the Connecticut River.

Acidic riverside seep at Garvins Falls on the Merrimack River in Concord.

Good Examples of River Channel Natural Communities

Hydric to Mesic Low Channel Communities: Pemigewasset River (Woodstock to Holderness), Saco River (Conway), Connecticut River (Lebanon to Cornish), Swift River (Albany), Peabody River (Gorham), Dead Diamond River (Second College Grant), Wild Ammonoosuc River (Bath), and Lamprey River (Epping to Newmarket).

Dry Low Channel Communities: Pemigewasset River (Woodstock to Holderness), Saco River (Conway), Connecticut River (Lebanon to Cornish), Swift River (Albany), Peabody River (Gorham), Dead Diamond River (Second College Grant), and the Wild Ammonoosuc River (Bath). Boulder - cobble river channels may be found at most of these sites, and are common along nearly all high-gradient streams, abundant in mountainous areas of the state.

Herbaceous High Channel and Floodplain Communities: Dry river bluffs occur on the Merrimack River (Franklin to Concord) and the Saco River (Conway). Riverwash plain and dunes occur only on the Merrimack River (Canterbury). Good examples of other communities in this category occur along the Merrimack River (Franklin to Manchester), Blackwater River (Salisbury to Webster), Contoocook River (Hillsboro to Concord), Piscataquog River (New Boston to Goffstown), Ashuelot River (Keene to Winchester), Lamprey River (Epping), Indian Stream (Pittsburg), and Dead Diamond River (Second College Grant).

Shrub High Channel and Floodplain Communities: Blackwater River (Salisbury to Webster), Contoocook River (Antrim to Hillsboro), Ashuelot River (Surry to Keene), Soucook River (Loudon), Exeter River (Exeter), Lamprey River (Epping), West Branch River (Ossipee), Upper Ammonoosuc River (Berlin), Indian Stream (Pittsburg), and Dead Diamond River (Second College Grant).

Riverbank Outcrops and Riverside Seep Communities: Good examples of all outcrop and seep communities are scattered along the middle third of the Connecticut River (Woodstock to Claremont). Acidic riverbank outcrops and acidic riverside seeps occur along the Ammonoosuc and Pemigewasset Rivers in the White Mountains, and Garvins Falls on the Merrimack River (Concord).

CHARACTERISTIC PLANTS OF
SELECTED LOW CHANNEL NATURAL COMMUNITIES

A = Dwarf cherry river channel   D = Willow low riverbank
B = Cobble - sand river channel   E = Twisted sedge low riverbank
C = Herbaceous sandy river channel   F = Mesic herbaceous river channel

COMMON NAME	SCIENTIFIC NAME	A	B	C	D	E	F
SHRUBS & SAPLINGS							
Dwarf cherry	*Prunus pumila*	●					
Eastern cottonwood	*Populus deltoides*	o	o				
Silver maple	*Acer saccharinum*	o	o	o			
Red maple	*Acer rubrum*	o	o	o			
Poison ivy	*Toxicodendron radicans*	o	o	o			
Dogwoods	*Cornus* spp.	o		o	o		
Alders	*Alnus* spp.	o	o		o		
Bristly dewberry	*Rubus hispidus*	o				o	
Willows	*Salix* spp.	o	o	o	●	o	
Meadowsweet	*Spiraea alba*				o	o	
HERBS							
Big bluestem	*Andropogon gerardii*	o	o				
Little bluestem	*Schizachyrium scoparium*	o	o				
Bedstraws	*Galium* spp.	o		o			
Twisted sedge	*Carex torta*	o	o	o		●	o
Prairie dogbane	*Apocynum cannabinum*	o	o		o	o	
Bent grasses	*Agrostis* spp.	o		o			o
Asters	*Aster* spp.		o				
Hawkweeds	*Hieracium* spp.		o				
Silverweed	*Argentina anserina*		o				
Lion's head*	*Physostegia virginiana**		o				
Bluegrasses	*Poa* spp.		o	o			
Horsetails	*Equisetum* spp.		o	o			
Spike-rushes	*Eleocharis* spp.			o			
Spotted touch-me-not	*Impatiens capensis*			o			
Joe-Pye weeds	*Eupatorium* spp.		o	o	o	o	
Smartweeds	*Polygonum* spp.		o	o	o		o
Goldenrods	*Solidago* spp.		o		o	o	
Panic grasses	*Panicum* spp.		o	o	o	o	o
Bluejoint	*Calamagrostis canadensis*	o	o	o	o	o	
Violets	*Viola* spp.			o	o	o	o
Swamp candles	*Lysimachia terrestris*			o	o	o	o
Mannagrasses	*Glyceria* spp.			o			o
Cut-grasses	*Leersia* spp.			o			o
Marsh fern	*Thelypteris palustris*			o			o
Common water hemlock	*Cicuta maculata*			o			o
Water parsnip	*Sium suave*			o			o
Sensitive fern	*Onoclea sensibilis*			o		o	o
Reed canary grass	*Phalaris arundinacea*					o	o
Water purslane	*Ludwigia palustris*						o

● = abundant to dominant   o = occasional or locally abundant   * = state threatened or endangered

CHARACTERISTIC PLANTS OF SELECTED HERBACEOUS OR SHRUB HIGH CHANNEL AND FLOODPLAIN NATURAL COMMUNITIES

A = Herbaceous riverbank/floodplain
B = Bluejoint - goldenrod - virgin's bower riverbank/floodplain
C = Meadowsweet alluvial thicket
D = Alder alluvial shrubland
E = Alder - dogwood - arrowwood alluvial thicket

COMMON NAME	SCIENTIFIC NAME	A	B	C	D	E
SHRUBS & SAPLINGS						
Box elder	Acer negundo	o				
Silky dogwood	Cornus amomum	o				•
Poison ivy	Toxicodendron radicans	o	o			o
Northern arrowwood	Viburnum dentatum	o				o
Nannyberry	Viburnum lentago	o				o
Meadowsweet	Spiraea alba	o		•	o	o
Rhodora	Rhododendron canadense			o		
Red osier dogwood	Cornus sericea	o			o	•
Speckled alder	Alnus incana	o			•	•
Willows	Salix spp.	o			o	o
HERBS						
Reed canary grass	Phalaris arundinacea	•				
Lake sedge	Carex lacustris	o				
Canada goldenrod	Solidago canadensis	•	o			
Tussock sedge	Carex stricta	o		o		
Fringed brome grass	Bromus ciliatus	•		o		
Bluejoint	Calamagrostis canadensis	•	•	o	o	o
Flat-topped goldenrod	Euthamia graminifolia	o	o		o	o
Mannagrasses	Glyceria spp.	o	o		o	o
Sedges	Carex spp.	o	o			o
Deertongue	Dichanthelium clandestinum	o				
Big bluestem	Andropogon gerardii	o				
Little bluestem	Schizachyrium scoparium	o				
Cut-grasses	Leersia spp.	o				
Swamp candles	Lysimachia terrestris	o	o	o	o	o
Sensitive fern	Onoclea sensibilis	o	o		o	o
Joe-Pye weeds	Eupatorium spp.	o	o		o	o
Bent grasses	Agrostis spp.	o	o			
Rough goldenrod	Solidago rugosa	o	•			
Panic grasses	Panicum spp.		o			
Smooth goldenrod	Solidago gigantea		•			
Bulrushes	Scirpus spp.	o	o			
Flat-topped white aster	Doellingeria umbellata		o		o	
Tall meadow-rue	Thalictrum pubescens		o		o	o
Spotted touch-me-not	Impatiens capensis		o		o	o
Virgin's bower	Clematis virginiana		•			o

• = abundant to dominant   o = occasional or locally abundant   * = state threatened or endangered

CHARACTERISTIC PLANTS OF
RIVERSIDE OUTCROPS, SEEPS, & BLUFFS

A = Acidic riverbank outcrop   D = Calcareous riverside seep
B = Circumneutral riverbank outcrop   E = Dry river bluff
C = Acidic riverside seep   F = Riverwash plain and dunes

COMMON NAME	SCIENTIFIC NAME	A	B	C	D	E	F
SHRUBS							
Dwarf bilberry*	Vaccinium cespitosum*	o					
Three-toothed cinquefoil	Sibbaldiopsis tridentata	o					
Leatherleaf	Chamaedaphne calyculata			o			
Large cranberry	Vaccinium macrocarpon			o			
Rhodora	Rhododendron canadense			o			
Golden heather*	Hudsonia ericoides*					o	
Creeping snowberry	Gaultheria hispidula			o			
Poison ivy	Toxicodendron radicans				o		
HERBS							
Path rush	Juncus tenuis	o					
Hawkweeds	Hieracium spp.	o					
Flat-topped goldenrod	Euthamia graminifolia	o					
Spiked false oats	Trisetum spicatum	o					
Sensitive fern	Onoclea sensibilis	o	o				
Goldenrods	Solidago spp.	o	o				
Violets	Viola spp.	o	o	o		o	
Allen's poverty oatgrass	Danthonia compressa	o				•	
Big bluestem	Andropogon gerardii	o	o	o	o	•	o
Little bluestem	Schizachyrium scoparium	o	o	o	o	•	o
New York aster	Symphyotrichum novi-belgii	o	o	o			
Tufted hairgrass	Deschampsia caespitosa	o			o		
Harebell	Campanula rotundifolia		o				
Dwarf ragwort*	Packera paupercula*		o				
Rusty woodsia	Woodsia ilvensis		o				
Christmas fern	Polystichum acrostichoides		o				
Siberian chives*	Allium schoenoprasum*		o				
Jesup's milk vetch*	Astragalus robbinsii*		o				
Small-flowered gerardia	Agalinis paupercula			o			
Round-leaved sundew	Drosera rotundifolia			o			
Bluets	Houstonia caerulea			o	o		
Kalm's lobelia*	Lobelia kalmii*				o		
Sticky false asphodel*	Triantha glutinosa*				o		
Garber's sedge*	Carex garberi*				o		
Grass-of-parnassus*	Parnassia glauca*				o		
Wild lupine*	Lupinus perennis*					o	
Shaved sedge	Carex tonsa						o
Stiff-leaved aster	Ionactis linariifolius						o
NON-VASCULAR							
Peat mosses	Sphagnum spp.			o			
Other mosses				o			o

• = abundant to dominant   o = occasional or locally abundant   * = state threatened or endangered

Arching canopy of silver maples along the Merrimack River.

Floodplain Forests

Tall, arching tree canopies and lush herbaceous layers characterize flood-plain forests. This group of communities occurs on flat, regularly flooded valley bottomlands adjacent to rivers and streams. Above bankfull stage, rivers overtop their banks, inundating floodplain terraces, scouring plants, and depositing fresh sediment. The dynamic floodplain environment supports numerous plant species and an abundance of amphibians, reptiles, birds, and mammals. Floodplain forests are often part of a mosaic of floodplain natural communities including thickets, meadows, and oxbow marshes and ponds. Some New Hampshire floodplain forest types are inherently rare due to the state's location at the edge of the geographic range of characteristic canopy species, such as swamp white oak, river birch, and sycamore.

Floodplain forest communities have a distinct structure, and contain a specialized suite of flood-tolerant species. Canopies are typically tall and closed, though less commonly they have a more open woodland structure. Silver maple, eastern cottonwood, and box elder are examples of trees that, in New Hampshire, grow only in floodplain forests. Vines are common in floodplains, but woody shrubs and tree saplings are sparse or absent, giving floodplain forests an open appearance. Tall ferns and broad-leaved herbs contribute to a dense understory during the growing season.

Floodplains are the product of sediment erosion and deposition. River-banks erode, especially during flood events, transporting sediments down-stream. Sediment deposition occurs when water rises above the riverbanks and loses velocity as it spreads across floodplain terraces. Coarser sediments fall out along edges of main channels and form sandy levees. Finer, sandy

Red maple floodplain forest along the Blackwater River in Salisbury.

or silty sediments settle on flat terraces and in various depressions behind the levees. Through this process, floodplains build and migrate downstream. The periodic influx of new sediment loads helps maintain fertile growing conditions.

The types and combinations of forested and open communities that occur in a given floodplain depend on differences in the timing, frequency, intensity, and duration of flooding, factors related to elevation and proximity to the river. Even slight changes in elevation result in different flood regimes and natural communities. Low floodplain forests are inundated every 1 to 3 years, whereas medium and high floodplain forests have longer flood-return intervals. Riverbank herb and shrub thicket communities often extend onto annually flooded portions of low floodplains. In many floodplain natural communities, grassy or sedgey areas occur on sandy levees. Oxbow marshes and vernal pools form in poorly drained depressions with silt and muck substrate. High terrace floodplains are largely isolated from flood dynamics, with flood intervals exceeding 100 years. These terraces support upland species that are intolerant of flooding and develop soils with distinct horizons, a characteristic absent in active floodplain soils. High floodplain forests were once extensive in New Hampshire, growing huge trees such as large "king pines" on productive, stone-free soils. Only remnant high floodplain forests remain today.

Watershed position, flow volume, and channel dynamics determine the distribution of floodplain forest communities. Although the dimensions of a river will vary along its length, the relationship of meander-length to river-width remains about ten to one. Floodplain forests are most extensive along low-gradient, meandering sections of large rivers. In broad lowland valleys of the Merrimack and Connecticut rivers, and on lower segments of their major tributaries, meander lengths are at least 0.5 mile and widths at least 250 feet. Silver maple dominates the low floodplains. On low-gradient sections higher in these tributary watersheds, and along smaller tributary rivers, meander lengths are 0.1 to 0.5 miles and widths only 50 to 250 feet. Red maple and other trees dominate floodplains along these rivers, sometimes with a thin border of silver maple growing right along the river. Flashy, braided rivers in mountainous areas and upper reaches of watersheds have narrow or discontinuous low floodplains, and sometimes more extensive medium to high floodplains. Meanders are about 0.5 miles in length and widths range from 200 to 250 feet. A mix of sugar maple, silver maple, ash, and fir dominate the natural communities in these settings.

Floodplain forest plants are flood-tolerant, and include many facultative wetland species. Adaptations to periodic inundation and physical disturbance include strong root systems such as the prolific rhizomes of nettles and sensitive fern; a perennial, clumped growth habit such as ostrich fern

Musclewood along a bank of the Soucook River.

and some grass species; and prolific production of seeds dispersed by wind, water, or animals. Most herbaceous plants grow rapidly and tall, attributes that help them compete for limited light. Some herbaceous plants sprout new growth from spreading underground rhizomes after burial by new sediments. Poison ivy, grapes, Virginia creeper, and Asiatic bittersweet vines grow rapidly above the herb layer to access sub-canopy and canopy light. By clinging to tree stems, vines gain protection from flood disturbance. Silver maple, box elder, eastern cottonwood, and red maple tolerate inundation and have shallow roots for oxygen uptake. In addition, these species sprout from roots or trunks near the soil surface in response to sediment deposition. Floodplain tree species are shade-intolerant, and require or favor bare mineral soil for seed germination.

New England floodplain forests are imperiled. Dam construction, agriculture, and development have displaced or modified floodplain forests and altered the natural processes that created and sustained them. Most remaining examples of these communities are smaller than 30 acres in size.

Floodplain Forests of Large Rivers

Silver maple and sugar maple dominate low floodplain forest communities along New Hampshire's large rivers. Other forest communities often occur adjacent to low floodplain communities on higher terraces. There are two silver maple and two sugar maple communities.

Two silver maple floodplain forest types collectively represent classic floodplain forests on large rivers in New Hampshire. Silver maple dominates the canopies of these forests, forming a tall, arching, cathedral-like ceiling above the level floodplain adjacent to the river channel. Shrubs are sparse,

Silver maple - false nettle - sensitive fern floodplain forest on the Saco River in Conway.

Silver maple - wood nettle - ostrich fern floodplain forest on the Connecticut River in Haverhill.

Sugar maple - ironwood - short husk floodplain forest on the Saco River in Bartlett.

and river grape and other vine species may be abundant along community edges and in canopy gaps. **Silver maple - false nettle - sensitive fern floodplain forest** is the most common silver maple community in New Hampshire, distinguished from its counterpart by a diverse and variable ground cover, abundant sensitive fern, and the presence of false nettle and common woodreed. Ostrich fern and wood nettle are sparse. This community is most common on the Merrimack River and medium-sized rivers such as the Ashuelot and Contoocook. Silver maple - false nettle - sensitive fern floodplain forest also occurs as a narrow floodplain border transitional to other floodplain forest types. **Silver maple - wood nettle - ostrich fern floodplain forest** is most common on the Connecticut River. Ostrich fern and northern lady fern dominate the understory, often with an abundance of wood nettle. Cottonwood, hackberry, and box elder trees occur in lesser numbers than silver maple. Graminoid and shrub cover is notably sparse. A rare plant called green dragon grows among ostrich fern at some sites.

Sugar maple, or a mix of sugar and silver maple, dominates two rare floodplain forest communities in New Hampshire. These communities occur along northern rivers with flashy flood regimes, and sometimes on floodplain terraces upslope of silver maple communities. Total species richness is high compared to other floodplain forest communities, although herb cover is sparse. **Sugar maple - silver maple - white ash floodplain forest** occurs primarily in the Saco and Androscoggin River watersheds. Sugar and silver maple are co-dominant, and white ash is also present. More upland forest herbs are present in this community than in the silver maple communities. Rich woods indicator species such as Jack-in-the-pulpit and blue cohosh are sometimes present. **Sugar maple - ironwood - short husk floodplain forests** have a dense tree canopy and the sparsest fern and herb layer of any floodplain forest community. Silver maples may be occasional immediately along the river but are otherwise low in abundance. This community is the predominant floodplain forest along upstream reaches of rivers that drain mountainous terrain, such as the Saco and Pemigewasset.

Floodplain Forests of Small Rivers

Red maple dominates or co-dominates floodplain forest communities along small rivers, where flood intensities and durations are reduced and flooding occurs earlier in the spring than on larger rivers. The shrub layer is often much denser in these communities than in silver and sugar maple floodplain forests. In each community, significant local variation in species com-

Red maple floodplain forest along the Lamprey River in Epping.

Balsam fir floodplain/silt plain on the Pine River in Ossipee.

Swamp white oak floodplain forest at Great Meadows in Kensington.

position is evident between lower- and higher-elevation terraces. Small river floodplain forests are common on lower sections of tributaries of the Connecticut and Merrimack rivers.

Red maple floodplain forests are common in southern and central New Hampshire, and absent from the northern part of the state. Soils range from somewhat poorly drained close to the river to moderately well drained at slightly higher elevations. Red maple dominates the canopy, with variable amounts of American elm, black cherry, and musclewood. The canopy structure ranges from forest to woodland. Winterberry, viburnums, and other shrubs form a low to moderately dense understory, and the herb layer is typically well developed and dominated by one or more ferns, such as sensitive fern, royal fern, or northern lady fern.

Balsam fir floodplain/silt plains occur in northern and central New Hampshire where balsam fir and other northern species of cold-climate, low-nutrient conditions are common. Balsam fir is interspersed with red maple, and occasionally silver maple and white pine. This community forms a forested to woodland zone above more frequently flooded alder thickets on some silty floodplains in the North Country.

Swamp white oak floodplain forests are state and regionally rare. In New Hampshire, the community only occurs in silty soils along drainageways within 30 miles of the coast. In addition to the dominant swamp white oak trees, other plants characteristic of this community include green ash, musclewood, drooping sedge, and loose-stemmed sedge. River birch, a state-rare tree, sometimes occurs in swamp white oak floodplain forests.

Sycamore floodplain forest is a regionally rare community type restricted in the state to floodplains along flashy sections of the Ashuelot River and a few other streams in the southern part of the state. Sycamore trees, which reach the northeastern limit of their range in southern New Hampshire, dominate the sparse to moderately well-developed tree canopy. The community often features a tall, well-developed musclewood shrub layer. Other characteristic species include red maple, sugar maple, American elm, and bitternut hickory. Butternut occurs less commonly. In areas where flooding is more frequent, the woody shrub and sapling layer may be absent, allowing for the growth of a tall, dense herbaceous layer dominated by jumpseed, ostrich fern, and bluejoint.

Good Examples of Floodplain Forest Natural Communities

Silver maple - wood nettle - ostrich fern floodplain forest: Connecticut River (Northumberland, Haverhill, and Walpole).

Silver maple - false nettle - sensitive fern floodplain forest: Merrimack River (Franklin to Concord), Contoocook River (Hillsboro to Con-

Sycamore floodplain forest along Great Brook in Walpole.

cord), Ashuelot River (Keene to Winchester), and Blackwater River (Salisbury).

Sugar maple - silver maple - white ash floodplain forest: Saco River (Conway) and Pemigewasset River (Woodstock to New Hampton).

Sugar maple - ironwood - short husk floodplain forest: Saco River (Bartlett to Conway).

Red maple floodplain forest: Ashuelot River (Surry), Contoocook River (Antrim), Blackwater River (Salisbury), Beaver Brook (Pelham), Lamprey River (Newmarket), the Exeter River (Exeter), Bearcamp River (Ossipee), Saco River (Conway), Swift River (Albany), and Indian Stream (Pittsburg).

Balsam fir floodplain/silt plain: Indian Stream (Pittsburg), Androscoggin River (Errol), Swift River (Albany), Pine River (Ossipee), and Big River (Barnstead).

Swamp white oak floodplain forest: Exeter River (Exeter), Powwow River (East Kingston to South Hampton), Tuttle Swamp (Newmarket), and Lamprey River (Newmarket).

Sycamore floodplain forest: Ashuelot River (Surry) and Great Brook (Walpole).

CHARACTERISTIC PLANTS OF SELECTED FLOODPLAIN FOREST NATURAL COMMUNITIES

A = Silver maple - wood nettle - ostrich fern floodplain forest
B = Silver maple - false nettle - sensitive fern floodplain forest
C = Sugar maple - silver maple - white ash floodplain forest
D = Sugar maple - ironwood - short husk floodplain forest
E = Red maple floodplain forest
F = Balsam fir floodplain/silt plain

COMMON NAME	SCIENTIFIC NAME	A	B	C	D	E	F
TREES							
Butternut	Juglans cinerea	o					
Hackberry*	Celtis occidentalis*	o					
Eastern cottonwood	Populus deltoides	o					
White ash	Fraxinus americana	o		•	o		
American elm	Ulmus americana	o	o	o		o	
Silver maple	Acer saccharinum	•	•	•	o		o
Sugar maple	Acer saccharum			•	•		
Black cherry	Prunus serotina			o		o	o
Ironwood	Ostrya virginiana				•		
Basswood	Tilia americana				o		
Red oak	Quercus rubra			o	o		
White pine	Pinus strobus			o	o		o
Red maple	Acer rubrum				o	•	•
Balsam fir	Abies balsamea						•
SHRUBS							
Poison ivy	Toxicodendron radicans		o	o	•		
Musclewood	Carpinus caroliniana					o	
Winterberry	Ilex verticillata					o	
Silky dogwood	Cornus amomum					o	
HERBS							
Wood nettle	Laportea canadensis	•					
Green dragon*	Arisaema dracontium*	o					
False nettle	Boehmeria cylindrica	o	•			o	
Ostrich fern	Matteuccia struthiopteris	•	o	o			o
Sensitive fern	Onoclea sensibilis	o	•	o		•	o
Northern lady fern	Athyrium filix-femina	o				•	o
Common woodreed	Cinna arundinacea	o	o				
Sessile-leaved bellwort	Uvularia sessilifolia			o	•		o
Northern short husk grass	Brachyelytrum septentrionale			o	•		o
Blue-stemmed goldenrod	Solidago caesia				•		
Wild sarsaparilla	Aralia nudicaulis				•		
Canadian germander*	Teucrium canadense var. vir*				o		
Cinnamon fern	Osmunda cinnamomea					o	
Royal fern	Osmunda regalis					•	o
White turtlehead	Chelone glabra					o	o
Bluejoint	Calamagrostis canadensis						o
Dwarf raspberry	Rubus pubescens						o

• = abundant to dominant   o = occasional or locally abundant   * = state threatened or endangered

Riparian Wildlife

River channel and floodplain communities support a diversity of wild-life species year-round, and serve as corridors for dispersal and seasonal movement. In spring, early plant growth in lowland riparian areas attracts deer, otter, and migrating birds seeking food. Seasonal floods fill floodplain sloughs and back channels creating fish breeding habitat. As floodwater recedes, fish trapped in pools become easy prey for raccoons, mink, and other predators. Leopard frogs inhabit floodplain marshes in the summer, but overwinter in permanent bodies of water too deep to freeze to the bottom. In late summer and fall, young turtles, birds, and raccoons disperse along rivers and streams in search of unoccupied habitat.

The state-threatened cobblestone tiger beetle inhabits cobble bars, laying its eggs in the muddy sand between the cobbles. Burrow-dwelling beetle larvae feed on passing insects and other arthropods. The more common bronze tiger beetle occupies sandier habitat. Both beetles are fierce predators of other insects.

Stoneflies and dragonflies use riverbank sediments to deposit larvae. Steep mud and sand riverbanks are important sites for the larvae of arrow clubtail dragonflies. They emerge in large numbers and fly along the river to feed, mate, and lay eggs. More gently sloped riverbanks and channel settings allow raccoons to fish, spotted sandpipers to probe for invertebrates, and deer to drink from the river. The variety of flowering herbaceous plants along riverbanks attracts butterfly species. Floods frequently undermine trees along the riverbank and cause them to fall into the river where they provide habitat for fish and other aquatic life.

Bank swallows and belted kingfishers dig nesting sites in sandy river bluffs. Bank swallows nest in colonies, whereas kingfishers are solitary nesters. Rivers provide ample food for the adults and young of both species—emerging insects for swallows, and insects and fish for kingfishers.

Wood turtles—a reptile whose populations are declining partly due to habitat loss—use a mix of forested, herbaceous, and open sandy areas within a quarter mile of cobbly or gravelly streams. In spring, they forage among silky dogwood, alder, and other shrubs for leaves, fruit, earthworms, and insects. Egg-laying occurs on warm, sandy banks that act as incubators. During the summer, wood turtles inhabit floodplain wetlands. In the winter, they hibernate in communal burrows on riverbanks or under woody debris.

Floodplain forests provide important habitat for wildlife species that are at the northern edge of their range in New Hampshire. Species that rely on particular plants associated with floodplain forests include the tawny emperor and the hickory hairstreak, dependent on hackberry and butternut, respectively. Floodplain forests are also important areas for birds.

Bald eagles roost in large floodplain trees, especially in winter. In spring, warblers migrating northward feed on insects among the emerging maple leaves and flowers. Other birds nest in silver maple floodplains, including warbling vireos, gray catbirds, song sparrows, and Baltimore orioles. Eastern wood-pewees, chestnut-sided warblers, and scarlet tanagers nest in red maple floodplain forests.

8 Seacoast

Winds, waves, and tides shape New Hampshire's 18 miles of seacoast. Despite the limited geographic extent of this region, it supports a great diversity of species and natural communities. The ocean is the driving force throughout, and salt significantly affects the three major habitat types discussed in this chapter: estuaries, sand dunes, and rocky shores.

At the peak of the last glacial advance, glaciers locked up enough of the world's water to lower the sea level. At the same time, the weight of the mile-thick ice sheet over New Hampshire depressed the land surface. A complex relationship developed between the global change in sea level and the isostatic rebound of land as the glaciers gradually melted. During this period, New Hampshire's coastline shifted both westward and eastward from its current position. The sediments that settled to the sea floor when the coastline lay inland of its current location have become the marine silt and clay deposits of today's Seacoast region.

In the wake of glacial retreat, the combined forces of rivers, waves, and ocean currents modified coastal topography by transporting, depositing, and eroding sediments. Sand moved along the shore, creating and eroding spits, bars, and beaches. Barrier islands and dunes formed where large amounts of sand collected and settled. About 4,000 years ago, the rate of sea level rise slowed enough for salt marshes to form behind barrier islands and in protected coves. As salt marsh vegetation died, the remains accumulated and formed a marine peat substrate that allowed the marshes to expand.

Today, the Seacoast region's rivers continue to deliver sediments to the sea, while waves and currents distribute these sediments along the coast.

PRECEDING PAGE: A variety of natural and human landscapes at New Hampshire's seacoast.

Great egrets forage in the salt marsh in Great Bay.

Waves are a dominant force in the seacoast environment.

Crescent-shaped sandy beaches, abundant in Seabrook and Hampton, alternate with rocky promontories. Extensive dune systems once existed landward of the sandy beaches, but coastal development has since displaced most of the state's dunes. Estuaries form where salt and fresh water mix in embayments protected from the sea. New Hampshire's largest estuary, Great Bay, is a complex mosaic of subtidal areas, tidal flats, and salt marshes. Rocky shores occur on wave-battered and salt-sprayed promontories, and are most abundant on the Isles of Shoals. Four of the nine islands in the Shoals archipelago belong to New Hampshire, while the rest are in Maine.

Variables that shape Seacoast communities include substrate type, moisture levels, frequencies of tidal inundation, salinities, and degrees of exposure to waves, wind, and blowing sand. Salt marshes and estuaries are among the most biologically productive systems on Earth, supporting a wide variety of plants and animals. Dunes and maritime rocky shores are less productive, yet also harbor specialized and distinctive life-forms. Collectively, these Seacoast habitats support a diverse suite of plants and animals, many of which are found nowhere else in the state.

Onshore winds drive coastal sands beyond beaches to form coastal dunes.

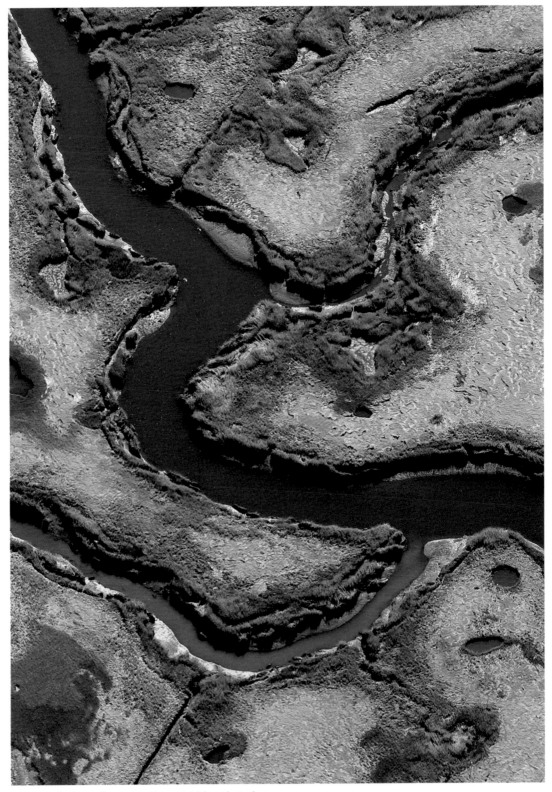

A mosaic of salt marsh communities and tidal creeks in the
Hampton Marsh estuary.

Estuaries

An estuary is a semi-enclosed body of water occurring where the lower course of a river nears the sea. Land barriers partially protect estuaries from high-energy wave action, and a combination of varying river volumes and daily tidal flows creates a mix of brackish, fresh, and salt water. Tides regularly flood and expose the intertidal zone, while a deeper subtidal zone is continuously submerged. Collectively, this productive ecosystem contains a high diversity of plants and animals adapted to dynamic ranges of tidal inundation and salinity.

Estuarine salinity is a function of mixing fresh and salt water. By definition, fresh water has a salinity of less than 0.5 parts per thousand (ppt), brackish water from 0.5 to 18 ppt, and salt water greater than 18 ppt. Estuarine salinity measures shift both daily and seasonally. Incoming tides force a wedge of dense salt water into estuarine bays and up tidal rivers. Conversely, fresh water flows down rivers and streams and into estuaries; because fresh water is less dense, it forms a lens above the salt water. Brackish conditions develop where the waters mix. Seasonally, estuarine salinity varies with changes in the volume of freshwater input, with the lowest salinites during spring when large volumes of freshwater are carried into estuaries as a result of snowmelt. By contrast, evaporation greatly increases salinity concentrations, particularly in pools that flood only infrequently.

Many factors and processes affect the distribution of estuarine communities, including the frequency of tidal flooding, storms, and ice scouring, salinity, soil oxygen, nutrient availability, interspecies interactions, sea-level

The Squamscott River is one of many rivers that feed freshwater into the Great Bay estuary. Estuaries form where fresh and salt water mix in protected embayments.

Salt-meadow cordgrass is the dominant plant in the **high salt marsh**, where it often forms wavy "cow-licks."

rise, and human land-use history. Furthermore, some of these factors are interrelated. The differences among species' tolerance ranges to these conditions result in distinct distribution patterns.

Flood frequency and salinity are two particularly important determinants of vegetation patterns in intertidal areas. Flood frequency depends on elevation. Ocean tides flood the lower areas of estuaries twice daily, inundating mudflats and low salt marshes. In contrast, the high salt marsh floods less than twice daily, on average. The distribution of cordgrasses and salt-marsh rush illustrates the importance of estuarine flood frequencies and salinity concentrations. Smooth cordgrass dominates low salt marshes, oxygenating its own root zone and tolerating regular tidal flooding. Salt-meadow cordgrass cannot tolerate the regular tidal flooding of the low marsh, but out-competes smooth cordgrass on high salt marsh, which is flooded less frequently. Along the upland edge of the high marsh, where flood frequency and salinity levels are lowest, salt-marsh rush out-competes salt-meadow cordgrass.

New England's salt marshes are physically different from other salt marshes along North America's Atlantic coast. Their fibrous marine peat contrasts with the more mineral soils of salt marshes to the north and south. Many plant species found in New Hampshire salt marshes are broadly distributed along the Atlantic coast, reaching their northern limits in the Canadian Maritime provinces. Although the flora does not differ significantly from region to region, soil differences could be important to other ecological features in these marshes. New Hampshire's estuaries provide habitat for nineteen rare plants and numerous uncommon species.

Humans historically dug ditches to drain salt marshes and increase the productivity of salt-meadow cordgrass and spike-grass. These plants were

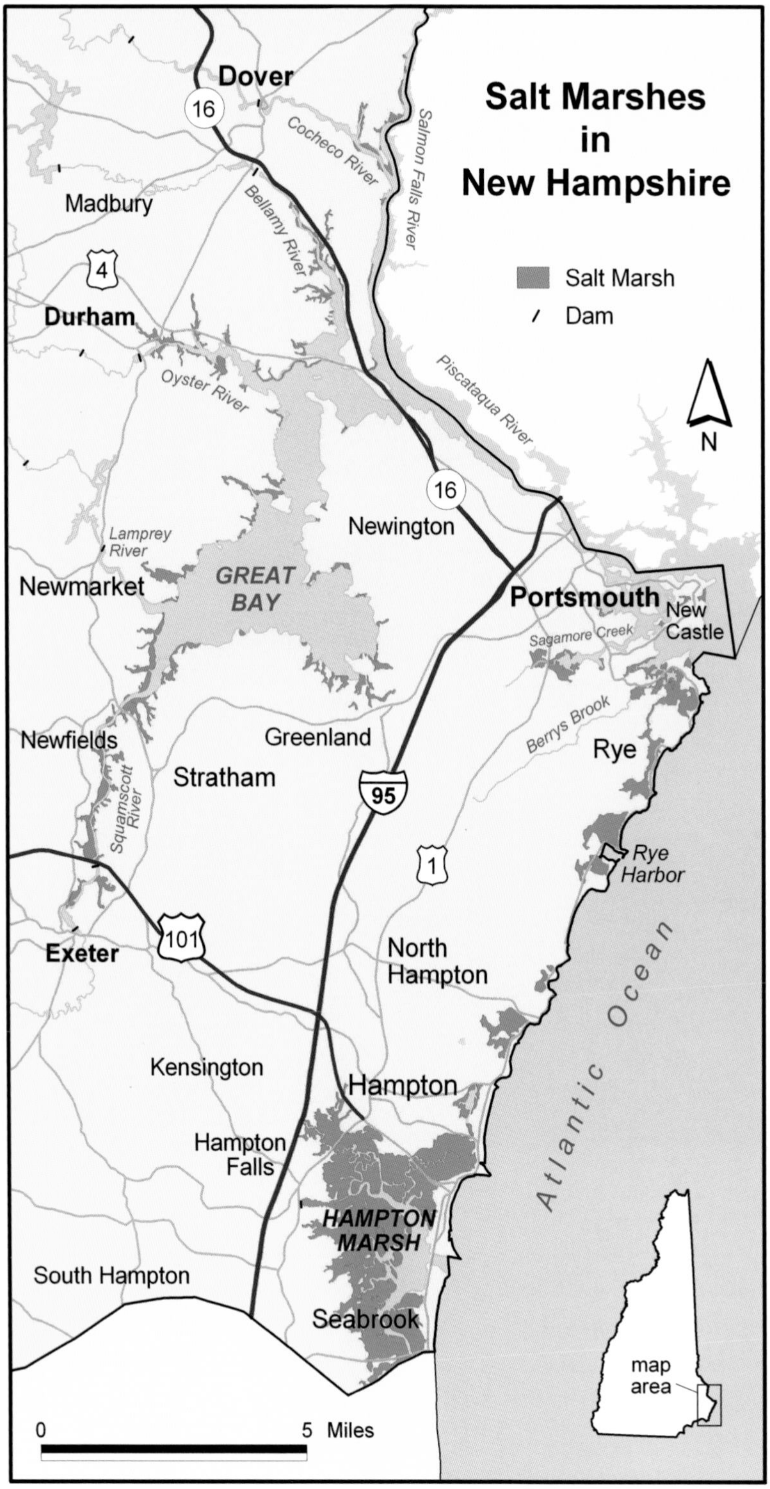
Salt Marshes in New Hampshire
Salt Marsh
Dam
N
Dover
16
Cocheco River
Salmon Falls River
Madbury
Bellamy River
4
Durham
Oyster River
Piscataqua River
Newington
Lamprey River
Newmarket
GREAT BAY
Portsmouth
New Castle
Sagamore Creek
Berrys Brook
Rye
Newfields
Greenland
Stratham
Squamscott River
95
1
Rye Harbor
Exeter
101
North Hampton
Atlantic Ocean
Kensington
Hampton
Hampton Falls
HAMPTON MARSH
South Hampton
Seabrook
map area
0
5 Miles

Turk's cap lily, an endangered plant, grows near the edge of its geographic range on estuarine borders in New Hampshire.

used for hay, pasture, and mulch. Salt haying declined in the early twentieth century, but ditching continued as a way to reduce mosquito populations. Ditching alters flood duration and water levels, which in turn affects community composition. For example, submerged aquatic plants such as widgeon grass, the state-threatened sago pondweed, and horned pondweed depend on the pools that ditching eliminates. Habitat loss and alteration also affects insects, mollusks, crustaceans, shorebirds, and waterfowl.

Another way humans have affected estuaries is by restricting tidal flow through the placement of fill and of inadequate culverts. Twenty percent of New Hampshire's salt marshes are degraded from tidal restrictions. Brackish or freshwater wetlands replace salt marshes where restrictions are severe. Other human-induced effects on estuaries include increased nutrient inputs, alteration of sediment supply, introduction of invasive plant species, and excess freshwater runoff. Mitigation efforts, like the one at Awcomin Marsh in Rye, have begun to restore the functions of healthy estuaries, which include buffering floods and providing wildlife habitat.

Salt Marshes

High salt marsh is the most common salt marsh community, accounting for more than 90 percent of the total salt marsh habitat in New Hampshire. This natural community is dominated by salt-meadow cordgrass, a slender grass with hollow stems often blown by winds into characteristic tufts and waves. Flooding of high salt marsh corresponds to lunar cycles; lower portions of

Many salt marshes were ditched and drained to control mosquitoes and increase salt hay production.

High salt marsh is the most common and extensive tidal marsh community in New Hampshire estuaries.

Low salt marsh forms narrow fringes between the high salt marsh and tidal flats of creeks or bays.

the community are flooded by higher-than-average tides, whereas the upper edge floods only during spring tides and storm surges. The upper edge frequently supports a broad diversity of plant species. Only one woody shrub species grows in New Hampshire salt marshes, the state-rare marsh elder. **Marsh elder shrubland**, which forms in just a few of the state's salt marshes, is a more common salt shrub community south of New Hampshire.

Low salt marsh occurs seaward of high salt marsh, and is flooded and exposed by the tides twice daily. Smooth cordgrass, usually much taller than salt-meadow cordgrass, dominates low salt marshes. The term "low" refers to the elevation of the marsh surface, not the height of the vegetation, though smooth cordgrass does have a low form that grows in salt pannes on the high marsh. This community often forms a narrow band between high marsh and tidal creeks or intertidal flats. Occasionally, smooth cordgrass occurs in small, isolated patches on the more frequently flooded upper mudflat. These cordgrass islands provide a substrate for aquatic mussels and facilitate further accretion of the salt marsh.

Salt pannes and pools occur as depressions embedded within salt and brackish marshes. Pools are deeper than pannes, and those on the high marsh are able to retain water and support marsh minnows between flooding by spring tides. Pannes and pools vary considerably in species richness and composition, depending on salinity, water levels, and substrate type. Salinity levels in pannes can rise significantly higher than seawater as a result of evaporation.

Salt pannes and pools are embedded in salt and brackish marshes and vary considerably in species composition, water level, and salinity.

ESTUARINE NATURAL COMMUNITIES

1. Upper reach of spring tide
2. Mean high tide
3. Mean sea level
4. Lower reach of spring tide

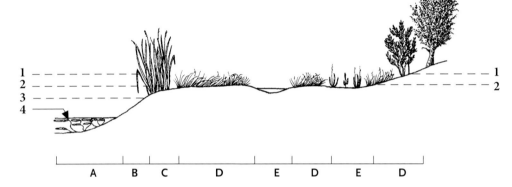

A. Eelgrass bed
Permanently inundated areas with eelgrass (other subtidal habitats possible)

B. Intertidal flat
Muddy, sparsely vegetated, flat or gently sloped lower intertidal areas; flooded twice daily

C. Low salt marsh
Narrow, middle intertidal zone between intertidal flats and high salt marsh; typically flooded twice daily

D. High salt marsh
Extensive peat flats in upper intertidal zone; flooded infrequently (typically less than twice daily)

E. Salt pannes and pools
Embedded in brackish and salt marshes, mostly in the irregularly flooded upper intertidal zone; deeper pool at left, forb panne at right

An idealized estuarine natural community sequence in a bay setting. Other community combinations are possible.

BRACKISH RIVERBANK NATURAL COMMUNITIES

1. Upper reach of spring tide
2. Mean high tide
3. Mean sea level
4. Lower reach of spring tide

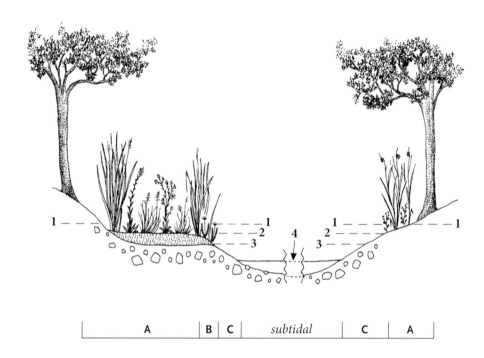

A. High brackish riverbank marsh

Irregularly flooded; typically less than twice daily; narrow zone on upper reaches of river; broad zone on lower reaches

B. Low brackish riverbank marsh

Regularly flooded; typically twice daily; narrow zone

C. Intertidal flat

Regularly flooded and irregularly exposed; muddy flats and gentle slopes

An idealized sequence of brackish riverbank natural communities along a tidal river in southeastern New Hampshire. Other community combinations are possible.

Brackish Marshes

Brackish marshes are transitional between freshwater and salt marsh communities. They occasionally form along the upland edges of tidal riverbanks salt marshes, where freshwater runoff or groundwater dilutes the salinity of the high marsh surface. This allows a mix of fresh and salt marsh species to coexist along with species restricted to brackish conditions. Most examples only flood during spring tides and storm surges. Dominant plant species and growth forms vary widely depending on local hydrology and salinity levels. Common species of brackish marshes are narrow-leaved cattail, stout bulrush, chaffy salt sedge, salt-meadow cordgrass, salt-marsh rush, and New York aster. Twelve rare plant species occur in this community.

Both **high brackish riverbank marshes** and **low brackish riverbank marshes** occur in fairly narrow zones along the upper reaches of estuarine rivers. High brackish riverbank marshes have a tidal flood frequency similar to high salt marshes, but have plants that require more freshwater influence, such as mudwort, eastern lilaeopsis, and false water pimpernel. Low brackish riverbank marshes flood twice daily, as do low salt marshes, but also support brackish-water tolerant species that are absent from more saline marshes.

Brackish marsh in Wallis Sands estuary.

A narrow band of vegetation on the banks of the Bellamy River at low tide. The **low brackish riverbank marsh** is indicated by the taller grass close to the river (smooth cordgrass); the less frequently flooded **high brackish riverbank marsh** by shorter grasses landward (common creeping bentgrass and salt-meadow cordgrass).

A wide example of **low brackish riverbank marsh** along the Cocheco River.

Intertidal Flats and Shores

Sparsely vegetated **intertidal flats**, often referred to as mudflats, are exposed at low tides. They occur between salt or brackish marshes and subtidal areas. There is little vascular plant growth here, but algae, worms, clams, snails, green crabs, and horseshoe crabs are all common. This community occupies a significant portion of New Hampshire's Great Bay estuary. Flats also extend up tidal rivers where they become progressively more brackish.

Intertidal rocky shores are most frequent on the open coast, but they also occur intermittently around estuarine embayments. Vascular plants are absent in the higher-elevation portions of this community, although cyano-

bacteria and lichens may be present. More frequently inundated parts of intertidal rocky shores support a great diversity of macroalgae. In the Great Bay estuary, seaweed species dominance shifts from green to red algae with increases in salinity levels. Great Bay supports disjunct populations of several seaweed species typically found in warmer climates. The rocky substrate also provides solid purchase for barnacles, periwinkles, and mussels.

Coastal shoreline strand/swales are sparsely vegetated communities that occur at the transition from estuarine to upland communities, particularly along muddy or rocky intertidal shores and the landward edge of salt marshes. Characteristic species include seabeach sand-spurrey, common glasswort, southern sea-blite, and sea lavender. The community often forms "wrack" along the strand line, heaps of decaying seaweed and herbaceous plants sheered off and deposited by tides and winter ice. Insects, arthropods, and aquatic life use the strand/swale for food and shelter.

Intertidal flats at Adams Point in Great Bay.

Subtidal

Subtidal areas are continuously submerged, below the reach of the lowest spring tides. They contain important habitat for oyster, eelgrass, and flounder populations, provide refuge for fish and invertebrates that retreat from exposed intertidal flats and estuarine marshes at low tide, and serve as

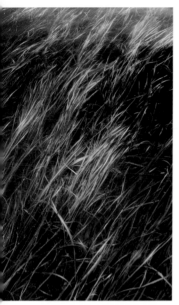

Eelgrass beds occur on muddy and sandy bottoms of Great Bay.

spawning and nursery areas for numerous species of aquatic animals. Eelgrass can form an extensive underwater meadow of **eelgrass bed**, a highly productive subtidal community that provides important habitat to many organisms. Eelgrass is a sensitive indicator of nitrogen loadings and is often used to monitor overall ecosystem health.

Good Examples of Estuarine Natural Communities

High salt marsh, **low salt marsh**, **brackish marsh**, and **coastal shoreline strand/swale** communities can be observed at Sagamore Creek at the Urban Forestry Center (Portsmouth), Rye Harbor State Park (Rye), Berry Brook at the western entrance of Odiorne Point State Park (Rye), Hampton Marsh (Hampton/Hampton Falls/Seabrook), Great Bay Discovery Center (Stratham), Lubberland Creek Preserve (Newmarket), Bellamy River Wildlife Sanctuary (Dover), Crommet Creek (Durham), and numerous other scattered locations around Great Bay (especially along creeks and by the mouths of rivers).

Marsh elder shrubland occurs at the Great Bay Discovery Center (Stratham).

High brackish riverbank marsh and **low brackish riverbank marsh** communities can be observed along the upper tidal reaches of the Lamprey, Cocheco, Bellamy, Squamscott, and Salmon Falls Rivers.

Eelgrass bed and **intertidal flat** communities can be observed in Great Bay.

CHARACTERISTIC PLANTS OF
SALT AND BRACKISH MARSH NATURAL COMMUNITIES

A = High brackish riverbank marsh   D = Low salt marsh
B = Brackish marsh   E = Low brackish riverbank marsh
C = High salt marsh   F = Salt pannes and pools

COMMON NAME	SCIENTIFIC NAME	A	B	C	D	E	F
SHRUBS							
Marsh elder*	Iva frutescens*			o			
HERBS							
Beach umbrella sedge	Cyperus filicinus	o					
Mudwort*	Limosella australis*	o				o	
Eastern lilaeopsis*	Lilaeopsis chinensis*	o				o	
Pygmy weed*	Crassula aquatica*	o				o	
False water pimpernel*	Samolus valerandi*	o				o	
Seaside crowfoot	Ranunculus cymbalaria	o				o	
Water parsnip	Sium suave	o	o				
Curly dock	Rumex crispus	o	o				
Purple loosestrife	Lythrum salicaria	o	o				
Creeping bentgrass	Agrostis stolonifera	●	●				
Salt-marsh rush	Juncus gerardii	o	o	o			
Red fescue	Festuca rubra	o	●	o			
New York aster	Symphyotrichum novi-belgii	o	●	o			
Seaside goldenrod	Solidago sempervirens	o	●	o			
Switch grass	Panicum virgatum	o	o	o			
Vanilla grass	Hierochloe odorata	o		o			
Fresh-water cordgrass	Spartina pectinata	o		o			
Small salt-marsh aster	Symphyotrichum subulatum	o		o			
Salt-marsh plantain	Plantago maritima	o		o			o
Spike grass	Distichlis spicata	o		●			o
Salt marsh gerardia*	Agalinis maritima*			o			o
Sea milkwort	Glaux maritima			o	o		o
Salt-meadow cordgrass	Spartina patens	●	●	●			o
Smooth cordgrass	Spartina alterniflora				●	●	o
Stout bulrush	Bolboschoenus robustus	o	●		o	●	o
Narrow-leaved cattail	Typha angustifolia	o	●				o
Salt-marsh water hemp	Amaranthus cannabinus	o			o	●	
Chaffy salt sedge	Carex paleacea	o	o				o
Three-square rush	Schoenoplectus pungens	o	o			o	o
Small spike-rush*	Eleocharis parvula*			o	o	o	o
Silverweed	Argentina egedii	o			o	o	o
Salt-loving spike-rush*	Eleocharis uniglumis*			o		o	
Sea lavender	Limonium carolinianum				o	o	o
Common glasswort	Salicornia depressa				o	o	o
Halberd-leaved orach	Atriplex prostrata				o	o	o
Seabeach sand spurrey	Spergularia salina				o	o	o
Sea blites	Suaeda spp.				o	o	o
Salt-marsh bulrush	Bolboschoenus maritimus					o	o
Widgeon-grass	Ruppia maritima						o
Sago pondweed*	Stuckenia pectinata*						o
Horned pondweed*	Zannichellia palustris*						o
Seaside arrowgrass	Triglochin maritimum						o

● = abundant to dominant   o = occasional or locally abundant   * = state threatened or endangered

Beach grass grassland on the foredune at Hampton Beach State Park.

Sand Dunes

Coastal sand dunes are a rare feature along New Hampshire's seacoast. Dunes form where currents, waves, and wind move sand landward into zones of accumulation, creating an extreme environment characterized by dry and shifting sands, onshore winds, and salt spray. The spits and barrier beaches protecting the estuary in Hampton Harbor represent the post-development remains of what was once an extensive dune system in this area.

The principal ingredient for dune formation is an abundant supply of sand, along with currents, waves, and wind to transport that sand. Conditions for dune formation became favorable in the New Hampshire coastal area about 4,000 years ago, when the rate of sea-level rise slowed. This resulted in less beach erosion, and allowed more sediment to accumulate along the shore. The abundant glacial deposits that wash into the sea from the Merrimack, Piscataqua, and other coastal rivers ensured a steady supply of sand.

The process of coastal dune formation is similar everywhere. Ocean waves and currents push sediments onto beaches, winnowing the sand from finer-grained materials that remain in suspension. The surf then drags the fine-grained particles back into the ocean. Large waves and storm surges push sand above the line of mean high tide, forming a beach ridge. Onshore winds then blow sand up and over the ridge, where it eddies into the lee side and settles. As winds blow sand farther inland, embryo dunes form behind plants or any other physical object. Dunes migrate as sand continues to shift from the windward to the lee side of the dune.

Seabrook Beach and the remnants of a coastal dune system. Dunes form where an abundant supply of sand is driven inland by onshore currents and winds.

Beach grass often sends up new shoots from horizontal stems called rhizomes.

Three recognizable zones form in most coastal dune systems: the foredune, interdune, and backdune. Plant height and total vegetation cover generally increase with distance from the sea. The foredune is the most exposed and dynamic zone, with actively shifting sand and a beach ridge that is shaped and reshaped by wind and storm waves. Beach grass dominates the foredune, with few other species present. The interdune is less exposed, and typically dominated by either beach grass or hairy hudsonia. It contains a greater diversity of species than the foredune. Only strong winds and storms are capable of transporting sand beyond the interdune into the backdune zone. Backdunes are typically the tallest, oldest, most stable, and most protected part of a dune system. Taller woody vegetation grows on the lee side of backdunes, forming maritime shrub thickets, wooded slopes, and wet hollows.

Hairy hudsonia, a state-threatened plant, has spreading rhizomes and short vertical branches that help it survive burial by shifting sand.

Dune plants are adapted to shifting sands. When beach grass is buried by sand, the plant grows new roots, which are better able to absorb nutrients than older roots. Beach grass can also sprout from root fragments exposed and torn from live plants by erosion. Many dune plants, including beach grass and hairy hudsonia, spread horizontally by means of underground rhizomes, a useful strategy in shifting sand. Because plants can transport nutrients and water garnered from one part of the plant to more-deprived segments, reliance on rhizomes is a common adaptation in environments where resources are in limited supply.

Dune plants are also adapted to salt spray, extreme temperatures, and dehydration. Thick, succulent leaves and waxy or hairy leaf surfaces help dune plants conserve water, and protect the plants from salt spray. Seaside goldenrod, bayberry, and hairy hudsonia have various combinations of these features. High temperatures and xeric conditions are greatest in the upper layer of sand in the foredune. To reduce heat stress and water loss, the leaves of beach grass roll up lengthwise in hot, dry weather, flattening out when conditions are cooler and moister. While the near-surface layer of sand is often dry, sand in subsurface layers remains moist and accessible to plants with deep roots. Beach plum and seaside goldenrod are examples of plants that access subsurface water through deep taproots.

More than one hundred native plants occur in New Hampshire's coastal dunes, including nine threatened or endangered plants. Many are restricted to or concentrated in coastal environments, and approach the northern limit of their range in New Hampshire or the Canadian Maritime provinces. Plants that fit this pattern include Gray's umbrella sedge, sea-beach sedge, sea-beach needle grass, seabeach pinweed, northern bayberry, beach grass, and hairy hudsonia.

A complex balance of shifting sands and stabilizing vegetation once

Northern bayberry has thick, waxy leaves that help reduce water loss in the hot, dry dune environment.

Sea-beach sedge is one of several plants restricted to sandy coastal habitats.

Beach grass grasslands dominate the foredune zone. Shifting sands, water stress, and salt spray are greater here than in interdune or backdune zones.

maintained an extensive, functional dune system from the state border in Seabrook north to Hampton Beach, similar to intact examples farther south in Massachusetts and the mid-Atlantic region. Today, heavy seacoast development has reduced this dune system to small, fragmented remnants.

Beach grass grasslands are most prominent on foredunes, but also occur in interdune and backdune areas. They occasionally occur as narrow strands of vegetation along the upper edge of beaches without dunes. Each of these settings has actively shifting sand. Beach grass, the dominant species, creates extensive colonies by spreading underground stems called rhizomes. Seaside goldenrod and hairy hudsonia may also be present. The low-growing hairy hudsonia dominates portions of the interdune in some areas, forming a **hudsonia maritime shrubland**. Associates of hairy hudsonia in this community include seaside goldenrod, beach grass, and beach pea.

Bayberry - beach plum maritime shrublands are moderate-height shrub zones found in the backdune area and in small, protected hollows in the interdune. Sandy soils here are more stable than those found in the foredune and exposed areas of the interdune, allowing coastal shrubs such as bayberry and beach plum to persist. **Maritime wooded dunes** are found in slightly more protected areas of the backdune. Black cherry and other trees grow up to 20 feet tall in the most protected areas, and form a shorter woodland canopy with a dense shrub layer in more exposed areas. Many of the woody species of the backdune, including serviceberry, black cherry, beach plum, juniper, and sassafras, produce large, edible fruits attractive to coastal birds. **Coastal interdunal marsh/swale** is a freshwater wetland community found in sandy depressions between dunes. Dominant plant species vary from swale to swale, but include large cranberry and shore rush.

Remnant Examples of Sand Dune Natural Communities

No intact dune systems remain in New Hampshire. However, remnant dune communities exist at The Sands and Seabrook Beach (Seabrook), Hampton Beach State Park and the mouth of Hampton Harbor inlet (Hampton), and Odiorne State Park (Rye). Good examples of intact sand dune natural communities occur south of the state line, on Plum Island in Massachusetts.

Hudsonia maritime shrublands are well developed in parts of the interdune zone that have active sand movement. Hairy hudsonia is the dominant low-growing shrub here.

Bayberry - beach plum maritime shrublands form on more stabilized portions of interdune or backdune zones.

Tall shrubs and gnarled black cherries dominate **maritime wooded dunes**.

COASTAL SAND DUNE NATURAL COMMUNITIES

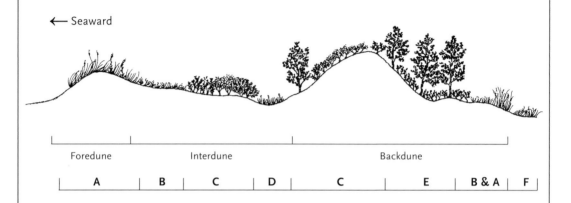

A. **Beach grass grassland**
Typically found on foredunes, heavily influenced by wind, salt spray, and storm waves: occasional in interdune and backdune zones

B. **Hudsonia maritime shrubland**
Shifting sand areas of interdune and backdune zones

C. **Bayberry - beach plum maritime shrubland**
Interdune and backdune zones where substrate is more stabilized than soils occurring in the foredune zone

D. **Coastal interdunal marsh/swale**
Low, wet, freshwater hollows of interdune or backdune zones

E. **Maritime wooded dune**
Backdune zones with the most stable soils

F. **Estuarine communities**
Dunes often transition to high salt marsh or other estuarine communities

An idealized coastal sand dune natural community sequence. Other community combinations are possible.

CHARACTERISTIC PLANTS OF SAND DUNE NATURAL COMMUNITIES

A = Beach grass grassland
B = Hudsonia maritime shrubland
C = Bayberry - beach plum maritime shrubland
D = Maritime wooded dune
E = Coastal interdunal marsh/swale
F = Coastal shoreline strand/swale

COMMON NAME	SCIENTIFIC NAME	A	B	C	D	E	F
TREES							
Black cherry	*Prunus serotina*			o	•		
Quaking aspen	*Populus tremuloides*			o	o		
SHRUBS							
Hairy hudsonia*	*Hudsonia tomentosa**	o	•	o			
Northern bayberry	*Myrica pensylvanica*			•	o		
Beach plum	*Prunus maritima*			•	o		
European barberry	*Berberis vulgaris*			o	o		
Virginia rose	*Rosa virginiana*			o	o		
Dwarf shadbush	*Amelanchier canadensis*			o	•		
Poison ivy	*Toxicodendron radicans*			o	o	o	
Purple chokeberry	*Photinia floribunda*			o	o	o	
Large cranberry	*Vaccinium macrocarpon*					•	
Winterberry	*Ilex verticillata*					o	
HERBS							
Sea-beach needle grass*	*Aristida tuberculosa**	o	o				
Gray's umbrella sedge*	*Cyperus grayi**	o	o				
Poverty oatgrass	*Danthonia spicata*	o	o				
Jointweed	*Polygonella articulata*	o	o				
Perennial umbrella sedge	*Cyperus lupulinus*	o	o				
Seabeach pinweed	*Lechea maritima*	o	o				
Smooth beach pea	*Lathyrus japonicus*	•	o				
Seaside goldenrod	*Solidago sempervirens*	•	o	o			
Beach grass*	*Ammophila breviligulata**	•	o	o			
Little bluestem	*Schizachyrium scoparium*	o	o	o			
Tall wormwood*	*Artemisia campestris**	o	o	o			
Yarrow	*Achillea millefolium*	o	o	o	o		
Canada mayflower	*Maianthemum canadense*				o		
Grove sandwort	*Moehringia lateriflora*				o		
Shore rush	*Juncus arcticus*					•	
Marsh St. John's wort	*Triadenum virginicum*					o	
Purple loosestrife	*Lythrum salicaria*					o	
Seabeach sand spurrey	*Spergularia salina*						o
Common glasswort	*Salicornia depressa*						o
Southern seablite	*Suaeda linearis*						o
Sea lavender	*Limonium carolinianum*						o
Seaside alkali grass	*Puccinellia maritima*						o
Salt-meadow cordgrass	*Spartina patens*						o
Sea rocket	*Cakile edentula*						o
Coast-blite goosefoot*	*Chenopodium rubrum**						o

• = abundant to dominant   o = occasional or locally abundant   * = state threatened or endangered

Seaside goldenrod on a rocky shore at the Isles of Shoals.

Rocky Shores

New Hampshire's maritime rocky shore communities occur on the Isles of Shoals and at scattered locations along the mainland. The small, offshore islands of the Isles of Shoals, some of which are in Maine, are considerably more exposed to the ocean—and battered by it—than nearby mainland shores. The sea has washed across the central valley of Appledore Island, largest of the Isles of Shoals, several times in the past century. Waves splashed the top of the White Island lighthouse, 80 feet above sea level, during a 2007 storm. The islands have a long history of human use and impacts to vegetation, but nesting seabirds and maritime forces are still important ecological factors affecting the distribution of plants and natural communities.

Although there is a long history of human use at the Isles of Shoals, the ocean remains the dominant force.

Tides, waves, salt spray, seabird guano, and the dehydrating effects of sun and wind are the dominant forces of the maritime environment. The result is a rocky landscape dominated by herbs and shrubs. Intertidal rocky shores ringing the islands are most exposed to tides and waves. Rocky barrens occur above the intertidal zone where exposure to storm waves and salt spray is still substantial. Shrub thickets occur on slightly higher ground which is even less exposed to ocean forces. Herbaceous meadows form where seabird activity is high. In New Hampshire, several of these natural communities are largely restricted to the Isles of Shoals. These communities are infrequent, small, or absent on the mainland due to human development, decreased exposure to the effects of the maritime environment, and the absence of large seabird nesting colonies.

More than 350 species of plants occur on New Hampshire's rocky shores, reflecting the varied conditions of rocky shore communities. More than a third of the plant species are non-native; these are abundant in areas affected by recent human disturbances and in larger seabird nesting sites. Native species are more abundant in areas away from humans and birds. Seven rare native species occur on rocky shores in New Hampshire. Three of these species—sea lyme grass, oyster plant, and Pennsylvania cinquefoil—are restricted in the state to the Isles of Shoals.

Sea lyme grass, shown here growing with flowering mustard plants, is a species primarily of northern maritime shores. In New Hampshire, it is found only on the Isles of Shoals.

Rocky shore natural communities reflect different degrees of exposure to tides, waves and storm surges, salt spray and salinity, and drought. Nesting birds are also influential. The scarcity or absence of trees is a striking feature of the Isles of Shoals. Here, all of the rocky shore communities occur together in a predictable pattern, extending landward from intertidal rocky zones to barrens, meadows, and thickets.

Intertidal rocky shores occur from the lowest shore exposed at low tide to the splash zone. They are subjected to waves and tides and, at low tide, sun and wind. These environmental factors have a strong influence on species composition and zonation. Vascular plants are absent. Numerous ma-

The sequence of communities on maritime islands—from intertidal areas, to rocky barrens, to shrub thickets—reflects a diminishing exposure to tides, waves, and salt spray.

Intertidal rocky shores are dominated by seaweeds and other marine algae.

rine algae species dominate the lowest shore, and cyanobacteria and lichens color the rocks in the splash zone. Marine algae at the Isles of Shoals and along the open coast include a diversity of brown algae, with fewer types of red and green algae.

Maritime rocky barrens occur above intertidal rocky shores. These exposed rocky barrens have a sparse herb cover and even fewer shrubs, which extend into protected crevices from meadows and thickets above. A combination of sparse soil, wind-driven salt spray, and seabird guano limit plant growth on the exposed bedrock. Small depressions within this community hold **brackish water pools**. Depending on salinity, fresh and brackish-water graminoids and forbs occur within and along the margins of these pools. Salinity fluctuates due to storm waves, salt spray, precipitation, and evaporation.

Maritime rocky barrens are sparsely vegetated outcrops battered by storms and salt spray.

Brackish water pools form in small depressions that are periodically inundated by salt water.

Maritime shrub thicket consists of dense shrubs with a low abundance of scattered trees and understory herbs. Some examples occur on thin-soiled ridgelines in exposed settings near rocky barrens, and feature only waist-high shrubs. In island interiors, shrub height and cover increase, and stunted trees and understory herbs also occur. Maritime thickets cover most of the interior on the larger islands.

Maritime meadows, characterized by forbs and graminoids, dominate upland habitats landward of maritime rocky barrens on most of the smaller islands of the Isles of Shoals, including Seaveys and White islands. Maritime meadows are invariably linked to seabird nesting colonies. Gulls and cormorants play a significant role in maintaining species composition and structure in this community. The seabirds produce large amounts of guano, which contains high concentrations of nitrogen and phosphorous. They also

Maritime shrub thickets are characterized by a dense tangle of berry-producing shrubs and vines, such as chokeberries, raspberries, blackberries, poison ivy, and Virginia creeper.

Maritime meadows are herb-dominated communities influenced by both the activity and the guano of nesting seabirds.

trample and pull vegetation. The combination of these stressors favors certain perennial grasses and annual forbs, while eliminating shrubs and other plants. Herb density varies greatly, being less dense in exposed areas with thin, dry soils, and more dense in protected areas with thicker, moister soils.

Maritime cobble beaches occupy coastal areas that are above tidal influence, but subject to storm surges and waves. Forbs and graminoids grow among a cobble or gravel substrate. Cobble beaches occur on the Isles of Shoals and are also associated with rocky promontories along the mainland coast, such as at Odiorne Point State Park.

Coastal salt pond marshes are a broadly defined community type with vegetation zones distributed along hydrologic and salinity gradients. These seasonally to semi-permanently flooded basins occur beyond the upper reach of spring tides, but are periodically infused with salt water during

Maritime cobble beach at Odiorne Point State Park. This community forms on sparsely vegetated upper shores.

storm events. Coastal salt pond marshes support mostly freshwater shallow emergent marsh species near their upper margins, and medium-depth emergent marsh species tolerant of wetter, more brackish conditions in lower areas. Salinities fluctuate in response to freshwater input, evaporation, and infusions of salt water. In New Hampshire, this community only occurs at Odiorne Point State Park, and that example has been altered by the establishment and high total cover of several invasive plant species.

Coastal salt pond marsh *(photo center)* at Odiorne Point State Park in Rye.

Coastal rocky headlands are restricted to windswept promontories and narrow peninsulas in Great Bay where salt spray is a major environmental factor. The community consists primarily of stunted eastern red cedar and black oak, with lesser quantities of other trees and shrubs, including black cherry, pitch pine, bayberry, and eastern shadbush. Soils are thin and acidic. Most examples are disturbed by humans, and contain numerous non-native species, such as common buckthorn, European barberry, and Kentucky bluegrass.

Coastal rocky headland on Thomas Point in Newington.

Good Examples of Rocky Shore Communities

Maritime rocky shore communities occur on the Isles of Shoals. **Intertidal rocky shores** and **maritime cobble beaches** can also be seen on the mainland at sites such as Fort Stark (New Castle) and Odiorne Point State Park (Rye). **Coastal rocky headlands** occur on several peninsulas in Great Bay. New Hampshire's only **coastal salt pond marsh** is at Odiorne Point State Park.

CHARACTERISTIC PLANTS OF SELECTED ROCKY SHORE NATURAL COMMUNITIES

A = Maritime shrub thicket
B = Maritime meadow
C = Maritime rocky barren
D = Maritime cobble beach
E = Intertidal rocky shore

COMMON NAME	SCIENTIFIC NAME	A	B	C	D	E
SHRUBS & SAPLINGS						
Black cherry	*Prunus serotina*	o				
Eastern shadbush	*Amelanchier canadensis*	o				
Black huckleberry	*Gaylussacia baccata*	o				
Northern bayberry	*Myrica pensylvanica*	o				
Virginia rose	*Rosa virginiana*	•				
Virginia creeper	*Parthenocissus quinquefolia*	•				
Purple chokeberry	*Photinia floribunda*	•				
Common blackberry	*Rubus allegheniensis*	•	o			
Red raspberry	*Rubus idaeus*	•	o			
Poison ivy	*Toxicodendron radicans*	•	o			
Nightshade	*Solanum dulcamara*	o	o	o	o	
HERBS						
Rhode Island bentgrass	*Agrostis capillaris*	o	o			
Red sorrel	*Rumex acetosella*	o	•			
Rough goldenrod	*Solidago rugosa*	o	•			
New York aster	*Symphyotrichum novi-belgii*	•	•	o		
Red fescue	*Festuca rubra*	•	•	o		
Yarrow	*Achillea millefolium*	•	•	o	•	
Quack grass	*Elymus repens*	•	•	o	•	
Seaside goldenrod	*Solidago sempervirens*		o	o	o	
Poor-man's pepper	*Lepidium virginicum*		o	o		
Birds' knotweed	*Polygonum aviculare*		o	o		
Common purslane	*Portulaca oleracea*		o	o		
Smartweeds	*Persicaria* spp.		•			
Common ragweed	*Ambrosia artemisiifolia*		•		•	
Wild radish	*Raphanus raphanistrum*		•		•	
Hedge mustards	*Sisymbrium* spp.				•	
Sea rocket	*Cakile edentula*				o	
Pigweed	*Chenopodium album*				o	
Oyster plant*	*Mertensia maritima**				o	
Beach pea	*Lathyrus japonicus*				o	
Sea lyme grass*	*Leymus mollis**				o	
NON-VASCULAR						
Algae *(seaweeds)*	*Fucus* spp. & several others					•

• = abundant to dominant   o = occasional or locally abundant   * = state threatened or endangered

Seacoast Wildlife

New Hampshire's estuaries are highly productive and diverse ecosystems, and serve as important habitat for many species of wildlife. Anadromous fish species such as Atlantic salmon, alewives, and shad pass through estuaries as they make their way from the open ocean to freshwater rivers and streams to spawn. In the winter, state-endangered bald eagles migrate south, where they roost in large pine trees on the edge of bays. Wading birds and waterfowl are among the many birds that feed on the plants, fish, insects, and marine invertebrates in estuaries. Seaside sparrows and salt-marsh sharp-tailed sparrows nest in cordgrass-dominated salt marshes and feed along the edges of ditches, pools, and salt pannes.

Low tides expose estuarine mudflats, revealing numerous species that inhabit mud and silt deposits. Common species here include mud snails, steamer and macoma clams, clam worms, oysters, blue mussels, and crustaceans. The horseshoe crab may be the most interesting mudflat species. This "living fossil" first appeared over 300 million years ago, prior to the dinosaurs. Horseshoe crabs can be seen in Great Bay during the summer months when females lay their eggs on the upper edges of mudflats and shorelines.

Sand dunes provide critical habitat to species such as the state-endangered and federally threatened piping plover. Piping plovers lay their eggs in the sand, often at the base of the foredunes where beach grass is sparse. Other species such as the semi-palmated plover, semi-palmated sandpiper, and sanderling use coastal sand dunes for resting and feeding during migration. Coastal dunes support numerous insects, some of which are restricted specifically to dune habitats. For example, flat-horned ant-beetles occur only on wooded backdunes, and eastern swift-horned ant-beetles burrow in the sand among beach grass on foredunes.

Many seabirds, waterfowl, and marine mammals breed on rocky shores, especially on the Isles of Shoals. The largest tern colony in the Gulf of Maine is located on Seaveys and White islands. The sparse vegetation and lack of mammalian predators make these islands prime locations for nesting terns. The restored colony now supports a large number of common terns, and smaller numbers of Arctic terns and federally endangered roseate terns. All three are rare in New Hampshire. The Isles of Shoals are also a staging area and migratory stopover for neotropical birds.

Marine wildlife along rocky shores includes barnacles, periwinkles, dog whelks, and blue mussels. Anemones, starfish, and crabs are common in tide pools. The common green crab is an invader from Europe, having arrived in the bilges of ships. They are voracious predators of mussels and clams, and indirectly affect native crabs. Japanese shore crabs are also a troublesome new invader.

Appendix 1

Global and State Rank Codes for Natural Communities

The New Hampshire Natural Heritage Bureau assigns conservation status ranks to natural communities based on assessments of rarity, trends, and threats. These ranks provide estimates of the risk of elimination. Use of standard criteria and rank definitions as defined by NatureServe and its member programs make these ranks comparable across community types and political boundaries.

Ranks are assigned for global status (G), which includes the entire range of a natural community, and statewide status (s). When a precise rank is unclear or the status of a natural community appears to fall between two ranks, a range may be used. For example, the range G4G5 could indicate that a community's global status rank is unclear but either 4 or 5, or that its rank is near the border between the two.

Rank	Description
1	**Critically Imperiled:** At very high risk of elimination due to extreme rarity (generally one to five occurrences), very steep declines, or other factors.
2	**Imperiled:** At high risk of elimination due to a very restricted range, very few examples (generally six to twenty occurrences), steep declines, or other factors.
3	**Vulnerable:** At moderate risk of elimination due to restricted range, relatively few examples (generally twenty-one to one hundred occurrences), or vulnerable to elimination because of other factors.
4	**Apparently Secure:** Uncommon but not rare; some cause for long-term concern due to declines or other factors.
5	**Secure:** Demonstrably common, widespread, and abundant.
U	**Uncertain:** There is evidence that a community may be eliminated throughout its range, but not enough evidence to state this with certainty. More information needed.
H	**Historical:** Known only from historical records, but with reasonable possibility of rediscovery.
X	**Eliminated:** Eliminated throughout its range, with no restoration potential due to extinction of dominant or characteristic taxa and/or elimination of the sites and disturbance factors on which the community type depends.

Appendix 2

Natural Communities of New Hampshire (with State Rank)

acidic northern white cedar swamp (s1)

acidic riverbank outcrop (s3)

acidic riverside seep (s1)

acidic *Sphagnum* forest seep (s3s4)

alder alluvial shrubland (s3)

alder - dogwood - arrowwood alluvial thicket (s4)

alder - lake sedge intermediate fen (s2s3)

alder seepage thicket (s3)

alder wooded fen (s3s4)

alpine heath snowbank (s1s2)

alpine herbaceous snowbank/rill (s1)

alpine ravine shrub thicket (s1s2)

alpine/subalpine bog (s1)

Appalachian oak - pine rocky ridge (s3)

Appalachian wooded talus (s1s2)

aquatic bed (s4s5)

Atlantic white cedar - giant rhododendron swamp (s1)

Atlantic white cedar - leatherleaf swamp (s1)

Atlantic white cedar - yellow birch - pepperbush swamp (s2)

balsam fir floodplain/silt plain (s2)

bayberry - beach plum maritime shrubland (s1)

bayonet rush emergent marsh (s2)

beach grass grassland (s1)

beech forest (s4)

Bigelow's sedge meadow (s1)

birch - mountain maple wooded talus (s3)

black gum - red maple basin swamp (s3)

black spruce - balsam fir krummholz (s2s3)

black spruce swamp (s3)

bluejoint - goldenrod - virgin's bower riverbank/floodplain (s3s4)

bog rosemary - sedge fen (s3)

boulder - cobble river channel (s3)

brackish marsh (s2s3)

brackish water pool (s2)

bulblet umbrella-sedge open sandy pond shore (s2)

buttonbush shrubland (s4)

calcareous riverside seep (s1)

calcareous sedge - moss fen (s2)

cattail marsh (s4)

chestnut oak forest/woodland (s1s2)

circumneutral - calcareous flark (s1)

circumneutral hardwood forest seep (s3)

circumneutral riverbank outcrop (s1)

circumneutral rocky ridge (s1)

coastal interdunal marsh/swale (s1)

coastal shoreline strand/swale (s2)

coastal rocky headland (s1)

coastal salt pond marsh (s1)

cobble - sand river channel (s3s4)

diapensia shrubland (s1)

dry Appalachian oak forest (s3)

dry red oak - white pine forest (s3s4)

dry river bluff (s3)

dwarf cherry river channel (s2)

eelgrass bed (s1)

emergent marsh (s5)

felsenmeer barren (s2)

floating marshy peat mat (s3s4)

hemlock - beech - oak - pine forest (s5)

hemlock - cinnamon fern forest (s4)

hemlock forest (s4)

hemlock - oak - northern hardwood forest (s4)

hemlock - spruce - northern hardwood forest (s3s4)

hemlock - white pine forest (s4)

herbaceous riverbank/floodplain (s2s4)

herbaceous seepage marsh (s3)

high brackish riverbank marsh (s1s2)

high-elevation balsam fir forest (s3s4)

high-elevation spruce - fir forest (s4)

high salt marsh (s3)

highbush blueberry - mountain holly wooded fen (s3s4)

highbush blueberry - sweet gale - meadowsweet shrub thicket (s4)

highbush blueberry - winterberry shrub thicket (s4)

hudsonia inland beach strand (s1)

hudsonia maritime shrubland (s1)

hudsonia - silverling river channel (s1)

inland Atlantic white cedar swamp (s1)

intertidal flat (s3)

intertidal rocky shore (s3)

jack pine rocky ridge (s1)

Labrador tea heath - krummholz (s2)

lake sedge seepage marsh (s3)

larch - mixed conifer swamp (s3)

large cranberry - short sedge moss lawn (s3)

leatherleaf - black spruce bog (s3)

leatherleaf - sheep laurel shrub bog (s2s3)

liverwort - horned bladderwort mud-bottom (s3)

low brackish riverbank marsh (s1s2)

low salt marsh (s3)

lowland spruce - fir forest (s3)

maritime cobble beach (s1)

maritime meadow (s1)

maritime rocky barren (s2)

maritime shrub thicket (s1)

maritime wooded dune (s1)

marsh elder shrubland (s1)

marshy moat (s4)

meadow beauty sand plain marsh (s1)

meadowsweet alluvial thicket (s3s4)

meadowsweet - robust graminoid sand plain marsh (s3s4)

mesic Appalachian oak - hickory forest (s2s3)

mesic herbaceous river channel (s4)

mixed alluvial shrubland (s4)

mixed pine - red oak woodland (s1s2)

mixed tall graminoid - scrub-shrub marsh (s4s5)

moist alpine herb - heath meadow (s1)

montane alder - heath shrub thicket (s1)

montane black spruce - red spruce forest (s1)

montane heath woodland (s2)

montane landslide barren and thicket (s3s4)

montane level fen/bog (s2)

montane lichen talus barren (s3)

montane sandy basin marsh (s1)

montane sandy pond shore (s1)

montane sloping fen (s1)

montane - subalpine acidic cliff (s5)

montane - subalpine circumneutral cliff (s2s3)

mountain holly - black spruce wooded fen (s3)

northern hardwood - black ash - conifer swamp (s2)

northern hardwood seepage forest (s3)

northern hardwood - spruce - fir forest (s4)

northern white cedar - balsam fir swamp (s2)

northern white cedar circumneutral string (s1)

northern white cedar forest/woodland (s1)

northern white cedar - hemlock swamp (s2)

northern white cedar seepage forest (s2)

oak - mountain laurel forest (s3)

pitch pine - Appalachian oak - heath forest (s1)

pitch pine - heath swamp (s1s2)

pitch pine - scrub oak woodland (s1s2)

pitch pine rocky ridge (s1)

red maple - black ash swamp (s3)

red maple - elm - ladyfern silt forest (s1s2)

red maple floodplain forest (s2s3)

red maple - lake sedge swamp (s3)

red maple - red oak - cinnamon fern forest (s3s4)

red maple - sensitive fern swamp (s3s4)

red maple - *Sphagnum* basin swamp (s4)

red oak - black birch wooded talus (s3s4)

red oak - ironwood - Pennsylvania sedge woodland (s2)

red oak - pine rocky ridge (s3s4)

red pine rocky ridge (s2)

red pine - white pine forest (s3)

red spruce - heath - cinquefoil rocky ridge (s3s4)

red spruce swamp (s3)

rich Appalachian oak rocky woods (s2)

rich mesic forest (s3)

rich red oak rocky woods (s2s3)

riverwash plain and dunes (s1)

riverweed river rapid (s2s3)

salt pannes and pools (s3)

seasonally flooded red maple swamp (s4s5)

seasonally flooded Atlantic white cedar swamp (s2)

sedge meadow marsh (s4)

sedge - rush - heath meadow (s1)

semi-rich mesic sugar maple forest (s3s4)

semi-rich oak - sugar maple forest (s2s3)

sharp-flowered mannagrass shallow peat marsh (s1)

sheep laurel - Labrador tea heath - krummholz (s2)

short graminoid - forb meadow marsh/mudflat (s4)

silver maple - false nettle - sensitive fern floodplain forest (s2)

silver maple - wood nettle - ostrich fern floodplain forest (s2)

Sphagnum rubellum - small cranberry moss carpet (s3)

spike-rush - floating-leaved aquatic mudflat marsh (s1)

spruce - moss wooded talus (s2s3)

subacid forest seep (s3s4)

subalpine cold-air talus shrubland (s1)

subalpine dwarf shrubland (s2)

subalpine rocky bald (s2)

subalpine sloping fen (s1)

sugar maple - beech - yellow birch forest (s5)

sugar maple - ironwood - short husk floodplain forest (s1)

sugar maple - silver maple - white ash floodplain forest (s1s2)

swamp white oak basin swamp (s1)

swamp white oak floodplain forest (s1)

sweet gale - alder shrub thicket (s3)

sweet gale - meadowsweet - tussock sedge fen (s4)

sweet pepperbush wooded fen (s2)

sycamore floodplain forest (s1)

tall graminoid meadow marsh (s4)

temperate acidic cliff (s4)

temperate circumneutral cliff (s2)

temperate lichen talus barren (s2s3)

three-way sedge - mannagrass mudflat marsh (s2s3)

twig-rush sandy turf pond shore (s1)

twisted sedge low riverbank (s3s4)

water lobelia aquatic sandy pond shore (s2)

water willow - *Sphagnum* fen (s3)

willow low riverbank (s3)

winterberry - cinnamon fern wooded fen (s4)

wire sedge - sweet gale fen (s3)

wooded subalpine bog/heath snowbank (s1s2)

Glossary

ablation till   Loose, permeable *glacial till* deposited from material in and on a glacier during its final melting (as opposed to compact *basal till* from material deposited under the ice). Ablation till often forms a more rolling, choppy landscape than *basal till*.

arboreal   Pertaining to trees; living in or among trees.

acidic   Used to describe water or soil with a *pH* less than 7.0. Although bedrock technically does not have a pH, the term "acidic" also can refer to bedrock with a total *silica* content greater than 65 percent, typically due to an abundance of *quartz, feldspar*, and mica. Acidic bedrock typically has low concentrations of *base cations* and yields nutrient-poor acidic soils with low pHs upon weathering. *Granite* is one example of an acidic rock. Compare *alkaline* or *basic, calcareous, circumneutral, intermediate, mafic,* and *subacid*.

adventitious   Growing in or from an unusual place, such as roots growing above ground, or roots that sprout from leaves.

A horizon   The dark-colored soil layer beneath the *O horizon*, where organic matter has been mixed with or incorporated into the mineral soil. An *E* or *B horizon* is usually found under the A horizon.

alkaline   Used to describe water or soils with a pH greater than 7 (synonymous with *basic*). Compare *acidic, calcareous, circumneutral, intermediate,* and *mafic*.

Alleghenian   A term applied to the forest distribution pattern that stretches from the glaciated northeastern United States and adjacent southern Canada to the Great Lakes region, with an extension into the central Appalachian Mountains. Representative species include red spruce, yellow birch, heartleaf birch, white pine, and hemlock.

alluvium   *Sand, silt,* and/or *clay* sediments deposited by moving water on land surfaces.

alpine   Refers to a largely treeless zone (also called tundra) found at high elevations, characterized by low temperatures, short summers, and high winds relative to surrounding lowlands. In New Hampshire, alpine areas are restricted to summits and ravines of the White Mountains above 4,900 feet (and occasionally on lower peaks and ridges), a few scattered high-elevation monadnocks in central and southern New Hampshire, and several cold microhabitats at lower elevations. Low-growing *graminoids*, shrubs, and herbs with northern or arctic centers of distribution are characteristic forms of vegetation. Compare *subalpine*.

backdune   In a coastal sand dune system, the dune that is farthest from the sea and the one least directly influenced by wave action and salt water. Compare *foredune* and *interdune*.

bankfull   The highest water level a *river channel* can reach prior to spilling over its banks onto its *floodplain*.

basal till   Base layer of compact *glacial till* deposited directly under the ice. Basal till is the most common soil parent material in New Hampshire, and like *ablation till*, it consists of an unsorted, unstratified mix of rock fragments, *sand, silt,* and *clay*. Basal till often forms relatively smooth landscapes, with frequent drainages, compared to the more rolling or choppy landscapes formed by *ablation till*.

base cations   Positively charged *ions*—including calcium, magnesium, sodium, and potassium—that generally have a higher concentration under *basic* (*alkaline* or *calcareous*) conditions.

basic   Precisely, any water or soil with a *pH* greater than 7.0; practically, applied to any soil with a pH greater than 7.4. Also loosely applied to an *igneous rock* that contains a relatively low amount of *silica* (45–52 percent) and a relatively high amount of *base cations* (although rocks do not technically have a pH). Compare *acidic, calcareous, circumneutral, intermediate,* and *mafic.*

B horizon   Mineral soil layer containing accumulations of nutrients, organic matter, and fine sediments leached from the *A* or *E horizon* above. Organic matter content is less pronounced than in the *A horizon* and occurs due to decomposition of roots or precipitation of organic coatings on mineral surfaces rather than by physical mixing of the soil.

biodiversity   The variety and variability of all living things. In its simplest and broadest terms, biodiversity refers to the full suite of life on Earth. More narrowly defined, it refers to a given area's full range of individual species and genetic diversity.

bog   Bogs are defined in both a narrow sense and a broad sense. The narrow hydrologic definition is a *peatland* whose only source of water is precipitation; the less restrictive definition (used in this book) is a nutrient-poor *peatland* with a *pH* less than 4. Bogs may receive minor inputs of *minerotrophic* water in the latter definition, but they remain floristically and trophically similar to strictly defined bogs. Compare *fen.*

boreal   Applied to a climate zone with short, warm summers and snowy winters. Also can refer to northern coniferous forests growing in a boreal climate, or individual species that have a boreal distribution. Boreal forest species in New Hampshire include balsam fir and quaking aspen. Compare *central hardwood, coastal plain,* and *transitional.*

bryophytes   Non-vascular, non-seedbearing plants, including mosses and liverworts.

calcareous   Applied to rock or soil containing calcite (calcium carbonate $CaCO_3$). When applied to rock, it implies that as much as 50 percent of the rock is calcium carbonate. When applied to soil, it implies that there is sufficient calcium carbonate (or other carbonates such as dolomite) to effervesce visibly or audibly when treated with cold, dilute hydrochloric acid (HCl). Calcareous soil generally has a *circumneutral* pH (in the 6s and 7s). Compare *acidic, alkaline, basic, circumneutral, intermediate,* and *mafic.*

calciphile   A plant species that grows best or is adapted to compete successfully in calcium-rich soils.

cations   Positively charged *ions.*

central hardwood   Applied to forest regions that have oaks, hickories, flowering dogwood, sassafras, and numerous other plant species found in the Appalachian states that reach their northern limit in or near southern New Hampshire. Compare *boreal, coastal plain,* and *transitional.*

C horizon   *Parent material* or soil that has been minimally modified by soil-forming processes.

circumneutral   Water or soil with a *pH* between 6.0 and 7.9. Compare *acidic, alkaline* or *basic, calcareous, intermediate,* and *mafic.*

cirque   Steep-walled, U-shaped ravines carved by *alpine* glaciers at high elevations in the mountains.

clay   A fine-grained, slow-draining mineral soil. The minerals that compose clay exhibit plasticity and harden when dried. Clays are typically associated with low-energy depositional environments such as large lakes. Compare *sand* and *silt*.

clonal growth   Plant reproduction by vegetative means such as spreading root systems.

coastal plain   A 10-mile to more than 100-mile wide biophysical region along the Atlantic and Gulf coasts, stretching from Texas to New England, characterized by certain plants that are restricted to or concentrated in this area. Evolutionarily, many of these plants have tropical origins. The coastal plain supports a high number of wetlands and is primarily composed of sedimentary rock. Compare *boreal*, *central hardwood*, and *transitional*.

colluvium   A combination of soil material and rock debris that has moved downslope by creep, slide, or local wash, and accumulated in lower landscape positions such as the bases of steep slopes or cliffs.

composite   A plant in the aster family (Asteraceae).

cosmopolitan distribution   Present over extensive areas of the globe; applicable to species such as field horsetail, lady fern, and many of our "weedy" plants.

delta   Flat, often fan-shaped landscape feature comprised of alluvial sediments deposited at the mouth of a river or stream.

discharge   The movement of water from *groundwater* reservoirs to the surface of the ground or to the atmosphere. Also, the volume of water flowing past a given point in a stream channel in a given period of time.

disjunct   Plant distributions that are notably distant from the primary range edge. For example, *alpine* areas in New Hampshire are said to be disjunct from the widespread ring of arctic flora found in the polar region to the north of the *boreal* forests.

drainage   Refers to the frequency and duration of periods of water saturation in soil. Drainage is affected by landscape position, soil texture, and groundwater fluctuations. A series of drainage classes (ranging from excessively well-drained to very poorly drained) have been described by the U.S. Department of Agriculture.

drumlins   Elliptical hills formed from *till* by the movement of glaciers, typically with a steep face on the side opposite the glacier's direction of travel. In New Hampshire, they are most common in the central and southern portions of the state. The bases of drumlins are often characterized by compact *basal till*.

dune   Low hill composed of *sand*. See *backdune*, *foredune*, and *interdune*.

E horizon   Soil layer beneath an *O horizon* or (less commonly) an *A horizon*. Mineral soil where most of the soil nutrients, organic matter, iron, and aluminum have been leached away to the *B horizon*. Usually indicates cool, moist, *acidic* environments. Fairly common in New Hampshire's forested soils from the White Mountains northward.

endemic   Refers to species that are restricted to a particular geographic region. For example, dwarf cinquefoil (*Potentilla robbinsiana*), only found at a few sites in the White Mountains, is considered endemic to New Hampshire.

esker   A long, narrow, steep-sided, and usually sinuous ridge of *sand* and gravel that has been deposited by meltwater from a glacier or ice sheet. They usually form either along the margins of a thinning glacier or along the pathways of drainage channels through the inside of a glacier. Subsidence of turbulent meltwater dumps the former streambed material in place, in the shape of the former channel.

estuarine   Referring to *estuaries*.

estuary   A coastal body of water and its adjacent wetland, semi-enclosed by land and with access to the sea, where ocean water is at least occasionally diluted by fresh water.

evapotranspiration   A collective term for the sum of water lost through evaporation from soils and bodies of water and transpiration through plants, usually expressed in comparison to rainfall amounts.

facultative   Capable of existing or functioning under multiple environmental conditions; not obligatory.

feldspar   The most common group of rock-forming minerals in the Earth's crust, usually found in *igneous rocks* such as *granite*, but present in *metamorphic* and *sedimentary rocks* as well. Feldspar crystals tend to be blocky, and white or pink in color.

felsenmeer   A German word meaning "sea of rocks," applied to extensive fields of large, angular, frost-cracked boulders found in *alpine* and arctic environments.

fen   A peat-accumulating wetland that develops where the water source has at least some mineral enrichment. Compare *bog* and *minerotrophic*.

floodplain   Nearly level land, consisting of sediments deposited by a stream or river along its borders and subject to periodic flooding. On average, the lowest floodplains flood every 1–3 years; this occurs when the river exceeds its *bankfull* discharge, the highest level of water the river channel can hold prior to spilling over its banks. River *terraces* with flood-return intervals of more than 100 years are referred to in this book simply as *terraces* (ancient floodplains); those with less than 100-year return intervals are referred to as floodplains.

fluvial   Applied to deposits of sediments transported and sorted by the action of moving water.

forb   Collective term for *herbaceous* plants, excluding *graminoids* (grasses and grass-like plants) and ferns.

foredune   In a coastal sand dune system, the dune nearest to the sea and the one most influenced by wave action and salt water. Compare *backdune* and *interdune*.

forest   Applied to natural communities with more than 60 percent cover by trees. Compare *woodland*.

glacial refugium   An ice-free geographic location in which vegetation survives during a glacial period (plural, *refugia*).

graminoid   Collective term for grasses and grass-like plants including sedges and rushes. Compare *forb*.

granite   A coarse crystalline *igneous rock* with at least 10 percent *quartz*, potassium *feldspar*, and plagioclase *feldspar* (calcium and sodium) in roughly equal portions as the essential minerals. "Dark minerals" (mica, pyroxene, and amphibole) comprise 3–10 percent. Weathers slowly to produce *acidic*, nutrient-poor soils.

groundwater   Water in the unblocked pores of soil sediments or fractures of bedrock below the *water table*.

headwall   The steepest, upper part of a *cirque*, often consisting of open rock slabs.

heath shrubs   Species in the family Ericaceae, including blueberries, cranberries, and crowberries. Heath shrubs, sometimes just called heaths, most commonly occur in *acidic*, nutrient-poor environments.

herbaceous   Non-woody vascular plants having no parts that persist above the ground after the growing season. Includes *forbs*, ferns, and *graminoids*.

hibernaculum   A shelter occupied during the winter by a dormant animal (a bat, for example).

hollow   A small, variably sized topographic depression (generally less than one to several feet in width and up to several feet deep), found in *peatlands* or uplands; may or may not support intermittent standing water. Hollows are separated from each other by *hummocks* and form as a result of a number of possible causes, such as windthrow in swamps and upland forests and differential decomposition and ice-wedging in some peatlands. Hollows are also called "pits," as in "pit-and-mound topography," particularly as applied to uplands. Compare *hummock*.

horizon   See *soil horizon*.

hummock   A mound in a *peatland* or upland community, up to several feet in height, and separated from other hummocks by *hollows* or "pits." Compare *hollow*.

humus   An upper soil *horizon* consisting of dark, well-decomposed organic material, resulting from the activity of biological organisms in the soil.

hydric   Pertaining to a condition of being inundated by water, or soils formed in poorly drained conditions (under the influence of permanent or intermittent inundation). Compare *mesic* and *xeric*.

ice-contact deposits   Variously sorted and *stratified* deposits formed from meltwater deposits in or at the edge of a thinning glacier (includes *kames*, *eskers*, and other undifferentiated deposits). Compare *outwash*, *talus*, and *till*.

igneous rocks   Rocks formed by the cooling and solidification of molten material. Compare *metamorphic rocks* and *sedimentary rocks*.

interdune   In a coastal sand dune system, dunes lying between the *foredune* and the *backdune*.

intermediate   Applied to *igneous rocks* that contain intermediate amounts of *silica* (50–60 percent), but generally less than 10 percent *quartz*. Intermediate rocks may produce *circumneutral* to *acidic* soils. Compare *acidic*, *alkaline*, *basic*, *calcareous*, *circumneutral*, and *mafic*. Also used in this book to refer to wetlands of intermediate nutrient status; in other words, between *oligotrophic* and strongly *minerotrophic*.

ions   Charged particles including *cations* (positive charge) and anions (negative charge). Nutrients become available to plants as elements and minerals in ionic form.

isostatic rebound   Rise in ground level associated with a landscape that has been depressed by the weight of a glacier.

kame   An irregular, short ridge or hill of *stratified* glacial deposits. Geologists have come to use this term more restrictively. *Ice-contact deposits* include various kame and kame-like deposits.

kettle hole   A depression formed from the melting of an ice block stranded in surrounding glacial sediment.

krummholz   Gnarled, stunted, and shrub-like conifer forest characteristic of timberline, generally less than six feet in height.

lagg   *Minerotrophic*, often moat-like, shallow water zone around the margin of a peatland, characterized variously by sedges, *forbs*, or tall shrubs, in combination with areas of open water. Lagg zones are influenced by *minerotrophic* upland runoff, are stagnant or slowly drained, and may have seasonally fluctuating water levels. They contrast with more interior portions of a peatland that are less influenced by upland runoff.

lakebed deposits   Fine sediments (consisting primarily of *clay* and *silt*) that settled in the quiet-water environments at the bottom of lakes.

levee   An elevated ridge of sediments formed parallel to a river by floodwaters, usually consisting of coarser sediments which settle due to the decreased water velocity away from the main river channel (finer particles are carried farther from the river).

lignin   A chemical compound found in the cell walls of all vascular plants, most notably as the fiber-like carbon polymer that gives strength to wood and allows trees to grow tall. One of the most abundant materials on Earth.

limestone   A *sedimentary calcareous* rock primarily composed of calcium carbonate. Produces calcium-rich soils with high *pH* and nutrient availability.

limiting nutrient   The nutrient that controls or limits the growth and productivity of plants, such that adding the nutrient would result in increased growth of some or all of the plants in a community.

loam   Soil composed of even concentrations of *sand, silt,* and *clay.* Loam soils retain nutrients and water easily, yet also have good infiltration and drainage. Considered ideal for gardening and agriculture.

mafic   Applied to minerals high in iron and magnesium or to *igneous rocks* relatively rich in such minerals. Compare *acidic, alkaline* or *basic, calcareous, circumneutral,* and *intermediate.*

mesic   Applied to moist, well-drained soils, intermediate between *xeric* and *hydric* soils. Also can refer to soils with mean annual temperatures between 47 and 59 degrees Fahrenheit at 20 inches below the surface.

metamorphic rocks   Rocks formed from recrystallization (without melting) of pre-existing rock (*igneous* or *sedimentary*) by high temperature and pressure.

microtopography   Small variations in the height and roughness of the ground surface, roughly at the scale of an individual organism.

mineralization   Decomposition of organic matter by microbes, converting bound, organic forms of minerals into ionic forms readily absorbed by plants.

minerotrophic   Refers to water or soils that have been enriched or contain nutrients important for plant growth; usually used in a relative or comparative sense.

montane   Refers to relatively cool, moist upland slopes below timberline in the mountains.

muck   A type of wetland soil composed of dark-colored, fine, well-decomposed and unrecognizable organic material, generally with a higher mineral sediment content than *peat.*

natural communities   Recurring assemblages of species found in particular physical environments. Each type is distinguished by three characteristics: (1) a definite plant species composition, (2) a consistent physical structure (such as *forest*, shrubland, or grassland); and, (3) a specific set of physical conditions (such as different combinations of nutrient, *drainage*, and climate conditions).

non-native   A species not indigenous to a particular area; alien or exotic. Invasive species are a subset of non-native species whose introduction to an ecosystem causes environmental harm.

nor'easter   Low-pressure storm system in which the strongest winds blow onshore from the northeast. They can originate any time of year along the east coast of North America as a result of converging (cold) polar and (warmer) Gulf Stream air masses, but can be most intense during the winter season.

O horizon   A soil layer found at the ground surface consisting of partially decomposed leaf litter and other organic matter, with little or no mineral soil component. Can be thick when conditions are either very wet and/or very cold and *acidic*.

obligate relationship   A relationship between two organisms that is necessary for the continued survival of at least one of the two.

oligotrophic   Nutrient-poor waters or soils with low primary *productivity*.

outwash   Glacial deposits that have been transported, reworked, sorted, and stratified by glacial meltwaters beyond the margin of the glacier. Compare *ice-contact deposits, talus,* and *till.*

oxbow   U-shaped portion of a river channel, occasionally abandoned as the river changes its course, leaving a depression in the *terrace*. Oxbow ponds can develop in these old river-channel depressions; they are persistently full of water and have species and processes similar to upland ponds.

palustrine   Referring to non-saline inland wetlands.

parent material   The original geologic material from which a soil is derived, usually referring to underlying sediment deposits or bedrock.

peat   Coarse, unconsolidated, largely undecomposed accumulations of organic matter formed under *hydric* conditions or in the presence of excess moisture. Compare *muck*.

peatland   Any *graminoid*, shrub, *forest*, or moss-dominated wetland formed on peat deposits.

pH   A measure of acidity or alkalinity, with values ranging from 0 to 14. Values below 7 are *acidic*, values above 7 are *alkaline* (or *basic*), and a pH of 7 is neutral. The pH of a solution can be calculated using the following equation, where $[H^+]$ represents the concentration of hydrogen *ions* in moles per liter: $pH = \log_{10}(1/[H^+])$. Because this equation defines a logarithmic scale, each unit pf pH represents a ten-fold change in acidity (a pH of 6, for example, is ten times more *acidic* than a pH of 7). See also *circumneutral* and *subacid*.

point bar   A low, crescent-shaped ridge of sand or gravel deposited in settings where flow velocities are low to moderate, such as along the inside bends of river meanders and at the downstream ends of islands.

poor   Infertile soils that have lower exchange capacity, base-saturation, *pH*, and/or nitrogen availability than *intermediate* or *rich* soils.

productivity   The biomass of vegetation generated over time in a given area (the total of which is called "gross primary productivity"), usually measured as above-ground biomass. Productivity varies greatly at both regional and local scales.

propagules   Parts of a plant, such as seeds, buds, rhizomes, or spores, which aid in dispersal and allow it to reproduce.

quartz   The second most common mineral in the earth's crust; a primary component of *granite*.

recharge   Movement of water from the surface into *groundwater* reservoirs.

rhizome   A horizontal, underground portion of plant stem, with shoots above and roots below, serving as a reproductive structure; not a true root (see *clonal growth*).

rich   Fertile soils that have higher exchange capacity, base-saturation, *pH*, and/or nitrogen availability than *intermediate* or *poor* soils. *Calciphytes* are characteristic indicators of rich soils.

rill   A small stream or rivulet, usually fast-flowing and narrow.

riparian   Term applied to the lands adjacent to and directly influenced by rivers and streams. *River channel* and *floodplain* environments are collectively called the riparian zone.

riverbank   The elevated ground bordering and containing a river.

river channel   Area between *riverbanks*, usually covered by water for at least a portion of the year.

riverine   Referring to rivers.

salt panne   A natural depression within a salt marsh inundated for longer periods than the surrounding marsh.

sand   The largest size class of particles that make up soil. Sand is gritty to the touch and individual grains are visible to the naked eye. Compare *clay* and *silt*.

sedimentary rocks   Rocks formed by consolidation of sediments deposited by wind or water. Compare *igneous rocks* and *metamorphic rocks*.

seepage   Lateral water flow through soil; transports nutrients to (and through) wetlands from soil and bedrock source areas. The water may emerge at the land surface where hydraulic pressure and an underlying impervious layer forces it from the subsurface, or it may remain below the surface where its presence is much more obscure. Many *fens* are influenced by some degree of groundwater seepage; **seeps** are small emergence zones on sloped uplands where seepage water emerges.

serotinous   Late or late-opening.

silica   Silicon dioxide (SiO_2). Silica commonly occurs as *sand* or *quartz*, and is a primary mineral component of *granite*.

silt   A soil particle intermediate in size between *sand* and *clay*. Silt is smooth and slippery when wet.

soil development   The process of physical and chemical alteration of *parent material* into soil under the influence of climate, hydrology, organisms, and time.

soil horizons   More or less horizontal layers of soil that have distinct physical and chemical characteristics resulting from soil-forming processes. The original *parent material*, climate, and the amount and quality of water that percolates through the soil profile are critical factors affecting the development of soil horizons. See also the definitions of individual horizons, including *O, A, E, B,* and *C*.

subacid   Used to describe water or soil with a *pH* from 5.0 to 5.9.

subalpine   Vegetation zone intermediate between high-elevation spruce - fir forests and alpine tundra, characterized by stunted spruce and fir trees (see *krummholz*) and a more limited number of *alpine* species than in higher-elevation regions. In New Hampshire, the subalpine zone generally occurs between 3,000 and 4,900 feet on exposed ridges and summits with thin soils. Some subalpine areas have been expanded by fires.

succession   Any sequential shift in vegetation that follows an environmental change (disturbance) or occurs as a result of some influence of the organisms themselves.

talus   A sloping mass of coarse rock debris, mostly stones and boulders, accumulated at the base of a cliff or slope. Sometimes referred to as scree. Compare *ice-contact deposits*, *outwash*, and *till*.

terrace   A nearly flat portion of a landscape which is terminated by a steep edge;

elevated portions of alluvial sediments, rock-cut benches, or other relatively flat features along a stream valley. Terraces can fall within the active *floodplain* and be flooded at some interval or may no longer flood due to downcutting of *river channels* and lower river volumes. Ancient *floodplain* terraces (beyond a 100-year flood-return interval) are referred to in this book simply as "terraces." Terraces within the 100-year floodplain are referred to either as "floodplains" or "floodplain terraces."

till   Deposited glacial material consisting of unsorted, unstratified *clay, sand, silt,* gravel, and rock fragments intermingled in any proportion. Compare *ice-contact deposits, outwash,* and *talus.*

tundra   Low-growing vegetation type characteristic of Arctic regions. *Alpine* vegetation is often referred to as tundra due to its similarity to Arctic tundra. See *alpine, subalpine.*

tussock   A compact tuft or clump of grass or sedge.

understory   Vegetation growing beneath a community's tallest woody plant stratum; the understory for a woodland or forest may include the shrub and herb layers; for medium to tall shrublands, just the herb layer.

water table   The upper surface or limit of *groundwater,* or that level below which the soil is saturated with water.

weathering   The physical and chemical process of breaking down *parent material* (including bedrock, *till*) and soil to forms more available to plants.

woodland   Applied to wooded communities with 25 to 60 percent cover by trees. Compare *forest.*

xeric   Characterized by a lack of moisture. Compare *hydric* and *mesic.*

References

Aber, J. D. and J. M. Melillo. 1991. *Terrestrial ecosystems*. Saunders College Publishing, Philadelphia.

Abrams, M. D. 2007. Tales from the blackgum, a consummate subordinate tree. *BioScience* 57: 347–59.

Anderson, D. S., and R. B. Davis. 1998. *The flora and plant communities of Maine peatlands*. Technical Bulletin 170, Maine Agricultural and Forest Experiment Station, University of Maine, Orono.

Andrus, R. E. 1980. *Sphagnaceae (peat moss family) of New York State*. Bulletin No. 442. New York State Museum, Albany.

Art, H. W. 1976. *Ecological studies of the Sunken Forest, Fire Island National Seashore, New York*. National Park Service Scientific Monograph No. 7.

Bailey, S. W. and J. W. Hornbeck. 1992. Lithological composition and rock weathering potential of forested, glacial till soils. U.S. Department of Agriculture Forest Service Research Paper NE-662.

Bailey, S. 2001. *A pilot study of the geology and ecology of cliff ecosystems in the White Mountains, New Hampshire*. U.S. Department of Agriculture Forest Service, Northeastern Research Station, Durham, N.H.

Baldwin, H. 1974. The flora of Mount Monadnock, New Hampshire. *Rhodora* 76: 205–28.

———. 1977. The induced timberline of Mount Monadnock, New Hampshire. *Bulletin of the Torrey Botanical Club* 104: 324–33.

Barrington, D. S., and C. A. Paris. 2007. Refugia and migration in the quaternary history of the New England flora. *Rhodora* 109: 369–86.

Bechtel, D. A., and D. D. Sperduto. 1998. *Floodplain forest natural communities along major rivers in New Hampshire*. New Hampshire Natural Heritage Inventory, Department of Resources & Economic Development, Concord, N.H.

Bedford, B. L., M. R. Walbridge, and A. Aldous. 1999. Patterns in nutrient availability and plant diversity of temperate North American wetlands. *Ecology*: 80: 2151–69.

Bellemere, J., G. Motzkin, and D. R. Foster. 2005. Rich mesic forests: Edaphic and physiographic drivers of community variation in western Massachusetts. *Rhodora* 107: 239–83.

Bertness, M. D. 1992. The ecology of a New England salt marsh. *American Scientist* 80: 260–68.

Billings, M. P. 1956. *The geology of New Hampshire*. State Planning and Development Commission. Reprinted 1980 by Division of Forests & Lands, Department of Resources & Economic Development. Concord, N.H.

Bliss, L.C. 1963. Alpine plant communities of the Presidential Range, New Hampshire. *Ecology* 44: 678–97.

———. 1963. *Alpine zone of the Presidential Range*. Privately published.

Bormann, F. H. and G. E. Likens. 1979. *Pattern and process in a forested ecosystem*. Springer-Verlag, New York.

Bormann, F. H., T. G. Siccama, G. E. Likens, and R. H. Whittaker. 1970. The Hubbard Brook ecosystem study: Composition and dynamics of the tree stratum. *Ecological Monographs* 40: 373–88.

Bornette, G., and C. Amoros. 1996. Disturbance regimes and vegetation dynamics: Role of floods in riverine wetlands. *Journal of Vegetation Science* 7: 615–22.

Boudette, E. L. 1990. The geology of New Hampshire, the Granite State. *Rocks and Minerals* 65: 306–12.

Brady, N. C. 1974. *The nature and property of soils*. MacMillan Publishing Co., New York.

Brinson, M. M. 1993. *A hydrogeomorphic classification for wetlands*. Technical Report WRP-DE-4. U.S. Army Corps of Engineers, Washington, D.C.

Brown, B. 1993. *A classification system of marine and estuarine habitats in Maine: An ecosystem approach to habitats. Part I: Benthic habitats*. Maine Natural Areas Program, Department of Economic and Community Development, Augusta, Me.

Carbonneau, L. E. 1981. Old-growth forest stands in New Hampshire: a preliminary investigation. Master's thesis, University of New Hampshire, Durham, N.H.

Chapman, D. H. 1974. New Hampshire's landscape: how it was formed. *New Hampshire Profiles* 23: 41–56.

Chapman, V. J. 1960. *Salt marshes and salt deserts of the world*. Interscience Publishers, New York.

Cleavitt, N. L. 2004. The bryophyte taxa of New Hampshire. *Evansia* 21: 49–75.

Cleavitt, N. L., R. E. Andrus, D. D. Sperduto, W. F. Nichols, and W. R. Town. 2001. Checklist of *Sphagnum* in New Hampshire. *Rhodora* 103: 245–62.

Cogbill, C. V. 1987. The boreal forests of New England. *Wildflower Notes* 2: 27–36.

———. 1994. *Vegetation of Franconia Ridge, New Hampshire: Historical ecology and management effects*. Report to U.S. Department of Agriculture Forest Service, submitted by New Hampshire Natural Heritage Inventory, Department of Resources & Economic Development, Concord, N.H.

Cogbill, C. V., J. Burk, and G. Motzkin. 2002. The forests of presettlement New England, USA: spatial and compositional patterns based on town proprietor surveys. *Journal of Biogeography* 29: 1279–304.

Cogbill, C. V., and P. S. White. 1991. The latitude-elevation relationship for spruce-fir forest at treeline along the Appalachian mountain chain. *Vegetatio* 94: 153–75.

Cowardin, L. M., V. Carter, F. C. Golet, and E. T. LaRoe. 1979. *Classification of wetlands and deepwater habitats of the United States*. U.S. Fish and Wildlife Service, Washington, D.C. FWS/OBS-79/31.

Crawford, M. M., C. E. Jeffree, and W. G. Rees. 2003. Paludification and forest retreat in northern oceanic environments. *Annals of Botany* 91: 213–26.

Crum, H. A. 1992. *A focus on peatlands and peat mosses*. University of Michigan Press, Ann Arbor.

Cutko, A. and T. J. Rawinski. 2007. Flora of northeastern vernal pools. In *Science and conservation of vernal pools in northeastern North America*, 71–104. CRC Press, Boca Raton, Fla.

Damman, A.W.H. and T. W. French. 1987. The ecology of peat bogs of the glaciated northeastern United States: A community profile. *Biological Report* 85(7.16). U.S. Fish and Wildlife Service, Washington, D.C.

Davis, M. B. 1976. Pleistocene biogeography of temperate deciduous forests. *Geoscience and Man* 13: 13–26.

———, ed. 1996. *Eastern old growth forests: Prospects for rediscovery and recovery*. Island Press, Washington, D.C.

Delcourt, H. R. and P. A. Delcourt. 1991. *Quaternary ecology: A paleoecological perspective*. Chapman & Hall, London.

Dunlop, D. A. and G. E. Crow. 1985. The vegetation and flora of the Seabrook dunes with special reference to rare plants. *Rhodora* 87: 471–86.

Dunlop, D. A., G. E. Crow, and T. J. Bertrand. 1983. *Coastal endangered plant inventory: A report on the Seabrook Dunes, its vegetation and flora.* Report prepared for the New Hampshire Office of State Planning by the Department of Botany and Plant Pathology and New Hampshire Agricultural Experiment Station, University of New Hampshire, Durham, N.H.

Engstrom, B. E. 1997. *Inventory and classification of natural communities along the Upper Saco River, New Hampshire.* New Hampshire Natural Heritage Inventory, Department of Resources & Economic Development, Concord, N.H.

Eyre, F. H., ed. 1980. *Forest cover types of the United States and Canada.* Society of American Foresters, Washington, D.C.

Fahey, T. J. 1976. The vegetation of a heath bald in Maine. *Bulletin of the Torrey Botanical Club* 103: 23–29.

Fernald, M. L. 1950. *Gray's manual of botany.* 8th ed. (corrected printing, 1970). Van Nostrand Co., New York.

Fincher, J. M. 1991. The relationship of soil-site factors to forest plant communities in the Green and White Mountain national forests. Master's thesis, University of New Hampshire, Durham, N.H.

Fincher, J., and M. L. Smith. 1994. A discriminate-function approach to ecological site classification in northern New England. U.S. Department of Agriculture Forest Service Research Paper NE-686.

Flaccus, E. 1959. Revegetation of landslides in the White Mountains. *Ecology* 40: 692–703.

Foster, D. R. 1992. Land use history (1730–1990) and vegetation dynamics in central New England, USA. *Journal of Ecology* 80: 753–72.

Foster, D. R., and E. R. Boose. 1992. Patterns of forest damage resulting from catastrophic wind in central New England, USA. *Journal of Ecology* 80: 79–98.

Foster, J. R., and W. A. Reiners. 1983. Vegetation patterns in a virgin subalpine forest at Crawford Notch, New Hampshire. *Bulletin of the Torrey Botanical Club* 110: 141–53.

Flora of North America Editorial Committee, eds. 1993+. *Flora of North America north of Mexico.* 14+ vols. New York and Oxford.

Fuller, J. L., D. R. Foster, J. S. McLachlan, and N. Drake. 1998. Impact of human activity on regional forest composition and dynamics in central New England. *Ecosystems* 1: 76–95.

Gawler, S., and A. Cutko. 2010. *Nautral Landscapes of Maine: A Guide to Natural Communities and Ecosystems.* Maine Natural Areas Program, Augusta, Me.

George, G. G. 1998. *Vascular plants of New Hampshire.* New Hampshire Natural Heritage Inventory, Department of Resources & Economic Development, Concord, N.H.

Gleason, H. A., and A. Cronquist. 1991. *Manual of vascular plants of northeastern United States and adjacent Canada.* 2nd ed. The New York Botanical Garden, Bronx, N.Y.

Goldthwait, J. W., L. Goldthwait, and R. P. Goldthwait. 1951. *The geology of New Hampshire. Part I: Surficial geology.* New Hampshire Department of Resources & Economic Development, Concord, N.H.

Golet, F. C., A.J.K. Calhoun, W. R. DeRagon, D. J. Lowry, and A. J. Gold. 1993. Ecology of red maple swamps in the glaciated Northeast: A community profile. *Biological Report* 12. U.S. Fish and Wildlife Service, Washington, D.C.

Grossman, D. H., D. Faber-Langendoen, A. S. Weakley, M. Anderson, P. Bourgeron, R. Crawford, K. Goodin, S. Landaal, K. Metzler, K.D. Patterson, M. Pyne, M. Reid, and L. Sneddon. 1998. *International classification of ecological communities: terrestrial vegetation of the United States. Volume I: The national vegetation classification system: development, status, and applications.* The Nature Conservancy: Arlington, Va.

Haines, A. (forthcoming). *Flora Novae Angliae.* New England Wild Flower Society, Framingham, Mass.

Harris, A. G., S. C. McMurray, P.W.C. Uhlig, J. K. Jeglum, R. F. Foster, and G. D. Racey. 1996. *Field guide to the wetland ecosystem classification for northwestern Ontario.* Ontario Ministry of Natural Resources, Northwest Science and Technology, Thunder Bay, Ont. Field Guide FG-01.

Henry, J. D. and J.M.A. Swan. 1974. Reconstructing forest history from live and dead plant material — an approach to the study of forest succession in southwest New Hampshire. *Ecology* 55: 772–83.

Johnson, C. 1985. *Bogs of the Northeast.* University Press of New England, Hanover, N.H.

Jump, A. S. 2005. Running to stand still: adaptation and the response of plants to rapid climate change. *Ecology Letters* 8: 1010–1029.

Junk, W. J., P. B. Bayley, and R. E. Sparks. 1989. The flood pulse concept in river-floodplain systems. In D. P. Dodge, ed., *Proceedings of the International Large River Symposium*, pp. 110–127. Canadian Special Publication of Fisheries and Aquatic Sciences 106.

Kenney, L. P., and M.R. Burne. 2000. *A field guide to the animals of vernal pools.* Massachusetts Division of Fisheries and Wildlife, Westborough, Mass.

Keys, J. E., and C. A. Carpenter. 1995. *Ecological units of the eastern United States: First approximation.* U.S. Department of Agriculture, Forest Service, Washington, D.C.

Larson, D.W., U. Matthes, and P.E. Kelly. 1999. Cliffs as natural refuges. *American Scientist* 87: 411–17.

———. 2000. *Cliff ecology: Pattern and process in cliff ecosystems.* Cambridge University Press, Cambridge.

Leak, W. B. 1979. *Why trees grow where they do in New Hampshire forests.* Report NE-INF-37-79, U.S. Department of Agriculture Forest Service, Northeastern Forest Experiment Station, Upper Darby, Pa.

———. 1982. Habitat mapping and interpretation in New England. U.S. Department of Agriculture Forest Service Research Paper NE-496.

Leak, W. B., and R. E. Graber. 1974. Forest vegetation related to elevation in the White Mountains of New Hampshire. U.S. Department of Agriculture Forest Service Research Paper NE–299.

Lull, H. W. 1968. *A forest atlas of the Northeast.* U.S. Department of Agriculture Forest Service, Northeastern Forest Experiment Station, Upper Darby, Pa.

Lyon, C. J., and W. A. Reiners. 1971. *Natural areas of New Hampshire suitable for ecological research.* Revised ed. Department of Biological Sciences publication no. 4. Dartmouth College, Hanover, N.H.

Maine Natural Areas Program. 1991. *Natural landscapes of Maine: A classification of ecosystems and natural communities.* Office of Comprehensive Planning, State House Station 130, Augusta, Me.

Malanson, G. P. 1992. *Riparian landscapes.* Cambridge University Press, Cambridge.

Malcolm, J. R., A. Markham, R. P. Neilson, and M. Garaci. 2002. Estimated migration rates under scenarios of global climate change. *Journal of Biogeography* 29: 835–849.

Marchand, P. J. 1987. *North woods*. Appalachian Mountain Club, Boston, Mass.

Mathieson, A. C., and C. A. Penniman. 1991. Floristic patterns and numerical classification of New England estuarine and open coastal seaweed populations. *Nova Hedwigia* 52: 453–85.

McDonnell, M. J. 1979. *The flora of Plum Island, Essex County, Massachusetts*. New Hampshire Agricultural Experiment Station Bulletin 513.

McQueen, C. B. 1990. *Field guide to the peat mosses of boreal North America*. University Press of New England, Hanover, N.H.

Metzler, K., and J. Barrett. 2003. *Vegetation classification for Connecticut*. State Geological and Natural History Survey of Connecticut, Department of Environmental Protection, Hartford, Conn.

Mitchell, C. C., and W. Niering. 1993. Vegetation change in a topogenic bog following beaver flooding. *Bulletin of the Torrey Botanical Club* 120: 136.

Moore, P. D., and D. J. Bellamy. 1974. *Peatlands*. Springer-Verlag, New York.

Motzkin, G. 1991. *Atlantic white cedar wetlands of Massachusetts*. Massachusetts Agricultural Experiment Station, University of Massachusetts. Research Bulletin 731.

National Wetlands Working Group. 1997. *The Canadian wetland classification system*. 2nd ed. Ed. B. G. Warner and C.D.A. Rubec. Wetlands Research Centre, University of Waterloo, Waterloo, Ontario.

Nichols, W. F., and V. C. Nichols. 2008. The land use history, flora, and natural communities of the Isles of Shoals, Rye, New Hampshire and Kittery, Maine. *Rhodora* 110: 245–95.

Nichols, W. F., D. D. Sperduto, D. A. Bechtel, and K. F. Crowley. 2000. *Floodplain forest natural communities along minor rivers and large streams in New Hampshire*. New Hampshire Natural Heritage Inventory, Department of Resources & Economic Development, Concord, N.H.

Nichols, W. F., D. D. Sperduto, and J. M. Hoy. 2001. *Open riparian communities and riparian complexes in New Hampshire*. New Hampshire Natural Heritage Inventory, Department of Resources & Economic Development, Concord, N.H.

Niering, W. A., and R. S. Warren. 1980. Vegetation patterns and processes in New England salt marshes. *BioScience* 30: 301–7.

Nixon, S. W. 1982. *The ecology of New England high salt marshes: A community profile*. U.S. Fish and Wildlife Service, U.S. Department of the Interior, Washington, D.C.

New Hampshire Natural Heritage Bureau. 2009. Biodiversity tracking and conservation system (BIOTICS) database. Concord, N.H.

Osgood, J. 1996. Contoocook River floodplain forest vegetation composition. Master's project, Antioch New England Graduate School. Keene, N.H.

Owen, C. R. 1999. Importance of hydrology, water quality, and disturbance to the northern basin marsh ecosystem of Grassy Pond, Litchfield, New Hampshire. Report submitted to The Nature Conservancy, Concord, N.H.

Parshall, T., and D. R. Foster. 2002. Fire on the New England landscape: regional and temporal variation, cultural and environmental controls. *Journal of Biogeography* 29: 1305–17.

Pease, A. S. 1964. *A flora of northern New Hampshire*. The New England Botanical Club, Cambridge, Mass.

Pielou, E. C. 1991. *After the ice age: The return of life to glaciated North America.* University of Chicago Press, Chicago.

Rawinski, T. J. 1984. *New England natural community classification.* Eastern Heritage Task Force, The Nature Conservancy, Boston, Mass.

——. 1985. Zonation and dynamics of riverwash Hudsonia barrens. Unpublished report. Eastern Heritage Task Force, The Nature Conservancy, Boston, Mass.

——. 1987. Pitch pine–scrub oak barrens of New Hampshire: endangered ecosystems. *Wildflower Notes* 2: 11–20.

Rawinski, T. J., L. A. Sneddon, and K. J. Metzler. 1989. The ecology of regularly flooded beach heather (*Hudsonia tomentosa*) vegetation along the Saco River: Community classification and interpretation. Unpublished report. Eastern Heritage Task Force, The Nature Conservancy, Boston, Mass.

Redfield, A. C. 1972. Development of a New England salt marsh. *Ecological Monographs* 42: 201–37.

Reimold, R. J. 1977. Mangals and salt marshes of Eastern United States. In V. J. Chapman, ed., *Wet coastal ecosystems*, pp. 157–66. Elsevier Scientific Publishing Co., Amsterdam.

Reiners, W. A., and G. E. Lang. 1979. Vegetation patterns and processes in the balsam fir zone, White Mountains, New Hampshire. *Ecology* 60: 403–17.

Reschke, C. 1990. *Ecological communities of New York State.* New York Natural Heritage Program, Latham, N.Y.

Rosgen, D. 1996. *Applied river morphology.* Wildland Hydrology Books, Pagosa Springs, Colo.

Royte, J. L., D. D. Sperduto, and J. P. Lortie. 1996. Botanical reconnaissance of Nancy Brook Research Natural Area. U.S. Department of Agriculture Forest Service Research Paper NE-216.

Schwarzman, B. 2002. *The nature of Cape Cod.* University Press of New England, Hanover, N.H.

Seymour, F. C. 1993. *The flora of New England: A manual for the identification of all vascular plants including ferns and fern allies growing without cultivation in New England.* Privately printed.

Short, F. T., ed. 1992. *The ecology of the Great Bay Estuary, New Hampshire and Maine: An estuarine profile and bibliography.* National Oceanographic and Atmospheric Administration, Coastal Ocean Program.

Siccama, T. G., F. H. Bormann, and G. E. Likens. 1970. The Hubbard Brook ecosystem study: productivity, nutrients, and phytosociology of the herbaceous layer. *Ecological Monographs* 40: 389–402.

Sorrie, B. A. 1994. Coastal plain ponds in New England. *Biological Conservation* 68: 225–33.

Sperduto, D. D. 1994. *Coastal plain pond shores and basin marshes in New Hampshire.* New Hampshire Natural Heritage Inventory, Department of Resources & Economic Development, Concord, N.H.

——. 2000. The vegetation of seasonally flooded sand plain wetlands of New Hampshire. Master's thesis, University of New Hampshire, Durham, N.H.

——. 2002. *A pilot study of the geology and ecology of cliff ecosystems in the White Mountains, New Hampshire: Vegetation.* New Hampshire Natural Heritage Inventory, Department of Resources & Economic Development, Concord, N.H.

——. 2005. *Natural community systems of New Hampshire.* New Hampshire Natural

Heritage Bureau, Department of Resources & Economic Development, Concord, N.H.

Sperduto, D. D., and S. Bailey. 2003. *Geology and ecology of cliff ecosystems in the White Mountains, New Hampshire: Second year report.* New Hampshire Natural Heritage Bureau, Department of Resources & Economic Development, Concord, N.H.

Sperduto, D. D., and C. V. Cogbill. 1999. *Alpine and subalpine vegetation of the White Mountains, New Hampshire.* New Hampshire Natural Heritage Bureau, Department of Resources & Economic Development, Concord, N.H.

Sperduto, D. D., and K. F. Crowley. 2002. *Atlantic white cedar in New England: Analysis and proposed classification.* New Hampshire Natural Heritage Inventory, Department of Resources & Economic Development, Concord, N.H.

———. 2002. *Floodplain forests in New England: Analysis and proposed classification.* New Hampshire Natural Heritage Inventory, Department of Resources & Economic Development, Concord, N.H.

Sperduto, D. D., and B. Engstrom. 1998. *Northern white cedar swamps of New Hampshire.* New Hampshire Natural Heritage Inventory, Department of Resources & Economic Development, Concord, N.H.

Sperduto, D. D., and A. Gilman. 1995. *Calcareous fens and riverside seeps in New Hampshire.* New Hampshire Natural Heritage Inventory, Department of Resources & Economic Development, Concord, N.H.

Sperduto, D. D., and S. Neid. 2003. *Exemplary bogs and fens in New Hampshire: Part II.* New Hampshire Natural Heritage Bureau, Department of Resources & Economic Development, Concord, N.H.

Sperduto, D. D., and W. Nichols. 2004. *Natural communities of New Hampshire.* New Hampshire Natural Heritage Bureau, Department of Resources & Economic Development, Concord, N.H.

Sperduto, D. D., W. F. Nichols, and N. Cleavitt. 2000. *Bogs and fens of New Hampshire.* New Hampshire Natural Heritage Inventory, Department of Resources & Economic Development, Concord, N.H.

Sperduto, D. D., W. F. Nichols, K. F. Crowley, and D. A. Bechtel. 2000. *Black gum (Nyssa sylvatica Marsh) in New Hampshire.* New Hampshire Natural Heritage Inventory, Department of Resources & Economic Development, Concord, N.H.

Sperduto, D. D., and N. Ritter. 1994. *Atlantic white cedar wetlands of New Hampshire.* New Hampshire Natural Heritage Inventory, Department of Resources & Economic Development, Concord, N.H.

Tappan, A., and M. Marchand. 2004. *Identification and documentation of vernal pools in New Hampshire.* 2nd ed. New Hampshire Fish and Game Department, Concord, N.H.

Taylor, J., T. D. Lee, and L. F. McCarthy, eds. 1996. *New Hampshire's living legacy: The biodiversity of the Granite State.* New Hampshire Fish and Game Department. Concord, N.H.

Thompson, L., and E. Sorenson. 2000. *Wetland, woodland, wildland: A guide to the natural communities of Vermont.* University Press of New England, Hanover, N.H.

Tiner, R. W. 1998. *In search of swampland: A wetland sourcebook and field guide.* Rutgers University Press, New Brunswick, N.J.

U.S. Department of Agriculture Natural Resources Conservation Service. Soil units digital data layer, revised 2002.

Vitousek, P. M., J. R. Gosz, C. C. Grier, J. M. Melillo, W. A. Reiners, and R. L. Todd. 1979. Nitrate losses from disturbed ecosystems. *Science* 204: 469–74.

Warren, R. S., and W.A. Niering. 1993. Vegetation change on a northeast tidal marsh: interaction of sea-level rise and marsh accretion. *Ecology* 74: 96–103.

Wessels, T. 1997. *Reading the forested landscape: A natural history of New England.* The Countryman Press, Woodstock, Vt.

Westveld, M. 1956. Natural forest vegetation zones of New England. *Journal of Forestry* 54: 332–38.

Whitlatch, R.B. 1982. *The ecology of New England tidal flats: A community profile.* U.S. Fish and Wildlife Service, Biological Services Program, Washington, D.C. FWS/OBS-81/01.

Whitney, G. G., and D. R. Foster. 1988. Overstory composition and age as determinants of the understory flora of woods of central New England. *Journal of Ecology* 76: 867–76.

Whitney, G. G., and R. E. Moeller. 1982. An analysis of the vegetation of Mt. Cardigan, New Hampshire: a rocky, subalpine New England summit. *Bulletin of the Torrey Botanical Club* 109: 177–88.

Woods, K. D. 1987. Northern hardwood forests in New England. *Wildflower Notes* 2: 2–10.

Worrall, J. J., and T. C. Harrington. Etiology of canopy gaps in spruce-fir forests at Crawford Notch, New Hampshire. *Canadian Journal of Forest Research* 18: 1463–69.

Worrall, J. J., T. D. Lee, and T. C. Harrington. 2005. Forest dynamics and agents that initiate and expand canopy gaps in *Picea-Abies* forests of Crawford Notch, New Hampshire, USA. *Journal of Ecology* 93: 178–90.

Zielinski, G. A., and B. D. Keim. 2003. *New England weather, New England climate.* University Press of New England, Hanover, N.H.

Illustration Credits

Scott W. Bailey: 62 (bottom)

Alan Briere: 121

Pete Bowman: 73 (bottom), 86 (bottom), 170, 178, 188 (bottom right), 199 (bottom), 270 (bottom)

Ben Kimball: xiv, 2, 7, 15, 23, 31, 33, 34 (top, bottom), 35 (all), 36, 39, 41 (right), 42, 43 (top, bottom), 46 (left top, left middle), 47 (middle), 48 (top, bottom), 51 (right), 53 (bottom), 54, 56 (bottom), 57 (bottom), 62 (top), 63, 66, 69, 72 (top, middle), 73 (top, middle), 79, 86 (top), 87, 88 (top, bottom), 89 (right), 95, 96, 102, 104, 105 (top), 106 (top), 107 (top, bottom), 108, 110, 111, 112, 114, 115, 116 (top), 117 (top, bottom), 118 (top, middle, bottom), 123, 124, 125 (middle), 126 (top), 130 (all), 135 (top, bottom), 137 (bottom), 147 (top), 152, 153 (left, right), 154 (top), 155 (bottom), 157 (top), 172 (top), 181 (left), 184, 187 (top left, top right, bottom right), 188 (bottom right), 189, 191 (top, bottom), 196 (left), 198 (middle, bottom), 199 (top), 201 (top left, bottom), 202 (bottom), 203, 207, 211 (right), 215, 216 (top, bottom), 217 (top), 218 (left, right), 221 (left), 222 (top left, bottom), 226, 227, 228, 229 (top), 232 (top, bottom), 233 (bottom), 237, 239, 240, 247, 249, 252 (top, bottom), 254 (top, middle), 258 (left top, bottom), 259 (bottom), 264, 265, 267, 268 (top), 269, 270 (top), 271, 275, 276, 277 (bottom), 280, 283 (top, bottom), 284 (top, bottom), 287, 289, 290, 292, 293, 294 (top), 296, 297 (top, middle, bottom), 302 (bottom), 304 (bottom), 305 (top, bottom)

Bill Nichols: 197 (top), 288 (bottom)

Mike Pelchat: 93

Dan Sperduto: 37, 38, 41 (left), 43 (middle), 44, 46 (left bottom, bottom), 47 (top, bottom), 50, 51 (left), 52, 53 (top, middle), 56 (top, middle), 57 (top), 58, 61, 68, 70 (top, bottom), 71 (top, bottom), 72 (bottom), 74, 76, 77, 78 (top, middle, bottom), 80, 81, 82 (left, right), 84, 85, 89 (left), 90, 98, 105 (bottom), 106 (bottom), 116 (bottom), 117 (middle), 120, 122, 125 (top, bottom), 126 (bottom), 127, 128, 132, 136 (top, bottom), 137 (top), 138, 141, 142, 147 (bottom), 150, 154 (bottom), 155 (top, middle), 156, 157 (bottom), 158, 162, 163, 164 (top, bottom), 165, 166 (top, bottom), 167, 168 (top, bottom), 169 (top, bottom), 171 (left, right), 172 (bottom), 179, 180, 181 (right), 185, 186 (left, right), 187 (bottom left), 188 (top left, bottom left), 190, 194, 195, 196 (right), 197 (middle, bottom), 198 (top), 200 (top, bottom), 201 (top right), 202 (top), 208, 210, 211 (left), 212, 214, 217 (bottom), 219 (left, right), 220 (top, bottom), 221 (right), 222 (top right), 229 (bottom), 230, 231, 232 (middle), 233 (top), 238, 241, 248, 250, 251, 253, 254 (bottom), 255 (top, bottom), 256 (top, bottom), 257, 258 (top, left middle), 259 (top, middle), 266, 268 (bottom), 277 (top), 278, 279, 282, 288 (top), 294 (bottom), 295 (top, bottom), 300, 301 (top, bottom), 302 (top), 303 (all), 304 (top), back cover (all)

Illustrations by Elizabeth Farnsworth

Maps by Ben Kimball

Index

For ease in use, only in-depth discussions or special-case examples of plant and place names are listed below; passing references in the Species Tables or Good Example sections are not included. **Boldface** text indicates natural community names. **Boldface** page numbers indicate the primary description(s) of the given community. Numbers in *italics* refer to illustrations or captions.